Photodynamic Therapy (3rd Edition)

Photodynamic Therapy (3rd Edition)

Guest Editor

Stefano Bacci

MDPI

Basel • Beijing • Wuhan • Barcelona • Belgrade • Novi Sad • Cluj • Manchester

Guest Editor
Stefano Bacci
Department of Biology
University of Florence
Florence
Italy

Editorial Office
MDPI AG
Grosspeteranlage 5
4052 Basel, Switzerland

This is a reprint of the Special Issue, published open access by the journal *Biomedicines* (ISSN 2227-9059), freely accessible at: https://www.mdpi.com/journal/biomedicines/special_issues/DB4X8X87W4.

For citation purposes, cite each article independently as indicated on the article page online and as indicated below:

Lastname, A.A.; Lastname, B.B. Article Title. *Journal Name* **Year**, *Volume Number*, Page Range.

ISBN 978-3-7258-4973-4 (Hbk)
ISBN 978-3-7258-4974-1 (PDF)
https://doi.org/10.3390/books978-3-7258-4974-1

Contents

About the Editor

Stefano Bacci

Stefano Bacci is a Professor of Cytology and Histology and Developmental Biology at the University of Florence and has resided, during his training, at the Schepens Eye Research Institute, Harvard Medical School, Boston (MA), USA. His main research concerns understanding the cellular mechanisms involved during wound healing in relation to the type of treatment used, including in relation to photodynamic therapy. He is the author of more than 100 articles within his research sector, including publications of high scientific impact.

biomedicines

MDPI

ditorial

Editorial: Photodynamic Therapy (3rd Edition)

tefano Bacci

Research Unit of Histology and Embryology, Department of Biology, University of Florence, Viale Pieraccini 6, 50139 Florence, Italy; stefano.bacci@unifi.it

Photodynamic therapy (PDT) was first demonstrated in 1903 by Von Tappeiner and Jesionek, combining light therapy with a photosensitizer and oxygen [1]. PDT is widely used in medical treatments for various human disorders, producing reactive oxygen species that destroy oncological tissue and pathogens [2].

Recent studies show that mild PDT can also have beneficial effects, such as wound healing, stress and pathogen resistance, and extended lifespan in animal models [3,4].

This editorial provides an overview of "Photodynamic Therapy (3rd Edition)", published in Biomedicine, highlighting recent contributions and advancements in PDT.

Research Articles

Gawecki et al. assessed the morphological and functional impacts of PDT in individuals with chronic central serous chorioretinopathy (CSCR), a retinal disorder characterized by the localized serous detachment of the macula [5]. The results indicated that a morphological response, characterized by the full clearance of subretinal fluid, was attained in 76.29% of patients, while an enhancement in best corrected visual acuity was seen in 77.53% of cases. All spectral domain optical coherence tomography assessments showed a substantial decrease after photodynamic therapy. Patients exhibiting intraretinal abnormalities or macular neovascularization had modest enhancements after PDT. A multivariate study indicated superior morphological outcomes correlated with younger age and male gender, as well as enhanced visual improvements in individuals devoid of intraretinal defects [6].

Regarding sensitizers [7], new boron dipyrromethene (BODIPY) compounds with a benzoxadiazole substituent have been recognized as promising sensitizers for PDT in cancer treatment. Porolnik et al. investigated the photochemical characteristics and cytotoxic activities of BODIPY in both normoxic and hypoxic environments. The dyes were analyzed via mass spectrometry and many nuclear magnetic resonance methods. In vitro investigations were performed on human ovarian cancer (A2780) and human breast adenocarcinoma (MDA-MB-231) cell lines. The incorporation of the benzoxadiazole moiety minimally influenced the location of the absorption maxima, but led to fluorescence quenching in comparison with mesophenyl-substituted analogs. Brominated and iodinated analogs exhibited significant light-induced cytotoxicity, characterized by low IC50 values ranging from 3.5 to 10.3 nM. Both compounds exhibited phototoxic action under hypoxic circumstances, with the iodinated BODIPY analog displaying the most significant cytotoxic impact. The research elucidates the benefits and possible limitations of BODIPY compounds, including heavy atoms and a benzoxadiazole moiety, as a valuable framework in medicinal chemistry for the development of novel photosensitizers [8].

The study of morphological changes in the skin during aging is a very current topic, and numerous research projects are being carried out. In general, with advancing age, collagen fibers diminish in thickness and transform in appearance, whilst elastic fibers exhibit fragmentation and a decrease in abundance. In the deep dermis, they gradually thicken due to distinct elastolysis mechanisms. The degradation and lysis of elastic fibers occur at a more rapid pace than in collagen fibers. With advancing age, the dermal vasculature diminishes owing to a decrease in both the quantity and size of vascular vessels, along with changes in the vascular wall constituents. These alterations progress until the vessel's operation terminates [9–12]. In this research, Notari et al. used a 675 nm laser wavelength on cultured human dermal fibroblasts to elucidate the signaling pathways involved in skin regeneration. The fibroblasts were irradiated 24 h post seeding, followed by the application of immunofluorescence microscopy and Western blotting. The findings indicate that laser therapy elicits substantial alterations in fibroblasts, influencing the cytoskeletal structure and the synthesis and reorganization of extracellular matrix (ECM) components. The cellular response to the therapy mostly involves paxillin-mediated signaling pathways. These modifications indicate that laser therapy may enhance the architecture and functionality of dermal tissue, with ramifications for addressing skin aging [13].

Microviscosity is an essential metric of the internal resistance to flow in a fluid, exemplified by the lipid bilayer of a cell membrane. It affects systems such as metabolism, signaling, and cellular motility [14–16]. Nonetheless, alterations in this biophysical parameter during PDT remain inadequately understood [14–16]. In this work, Shimolina et al. investigated alterations in the microviscosity of live HeLa Kyoto tumor cells under photodynamic therapy with KillerRed, a genetically encoded photosensitizer, within several cellular localizations. The results show that the nuclear localization of KillerRed resulted in a progressive reduction in microviscosity, while its membrane localization initially rose in the first minutes of PDT before subsequently declining. The findings indicate that membrane microviscosity plays a role in tumor cell responsiveness to photodynamic therapy, contingent upon the location of reactive oxygen species targeting a genetically encoded photosensitizer [17].

Cervical cancer ranks as the fourth most common malignancy among women, with approximately 600,000 new cases and 340,000 related deaths occurring globally each year [18]. Persistent infection with high-risk human papillomavirus is strongly associated with the development of cervical cancer. Wei et al. examined the effectiveness of minimally invasive 5-aminolevulinic acid (ALA) PDT in the treatment of cervical high-grade squamous intraepithelial lesions (HSIL/CIN2). The results indicated that smoking and sleep disturbances were independent risk factors for unsuccessful lesion regression during ALA-PDT. This indicates the need for the meticulous evaluation of ALA-PDT for individuals with these disorders [19].

Glioblastoma, a rare malignant brain tumor, has a poor prognosis, with a median overall survival of 14.6 months and a 5-year survival rate of 9.8%, despite standard treatments like maximal surgical resection, 60-Gy radiotherapy, and chemotherapy [20–22].

Saito et al. investigated the efficacy of interstitial photodynamic treatment (i-PDT) for malignant gliomas in the brain. C6 glioma cells were introduced into the basal ganglia of rats, followed by intraperitoneal administration of the second-generation photosensitizer, talaporfin sodium (TPS). A prototype optical cable was inserted into the tumor tissue, and semiconductor laser light was emitted into the tumor under different settings. The brain was excised 24 h post i-PDT and subjected to pathological analysis. Histological investigation revealed that tumor necrosis occurred proximate to the light source, but that apoptosis was elicited at a distance. Elevated energy levels led to tissue edema, and irradiation at 75 J/cm^2 was optimal for inducing apoptosis. The research developed an experimental i-PDT system using TPS, showing that tumor cell mortality is associated with light propagation [23].

Oral squamous-cell carcinoma and pancreatic cancer are lethal conditions that are becoming more prevalent and have low survival rates. In 2021, these two malignancies were responsible for 62,210 and 54,010 new cases, respectively [24,25]. The impact of a novel gel containing 5% v/v 5-ALA (ALAD-PDT) on human oral CAL-27 and pancreatic CAPAN-2 cancer cell lines was investigated in a study by D'Antonio et al. [26]. Flow cytometry demonstrated that the viability was greatly diminished at all concentrations, and that ALA-PDT induced significant apoptosis rates in both cancer cells. Both cell lines exhibited elevated levels of PpIX and ROS production. ALA-PDT has the potential to be administered as either a topical or intralesional therapy, which would enable the administration of modest dosages and reduce the likelihood of adverse effects. It has the potential to be a significant factor in the development of complex oral and pancreatic anticancer therapies [26].

Review

Bacterial infections and cancer are significant public health concerns on a global scale [27–29]. Antibiotics and corticosteroids are employed to treat bacterial infections, while chemotherapy medications, surgery, and radiotherapy are employed to combat cancer [27–29]. Resveratrol, a natural substance derived from Chinese herbal plants, has been employed to treat cancer and bacterial infections. It can also function as an adjuvant or photosensitizer in PDT. Resveratrol has been demonstrated to possess antibacterial and anticancer properties in studies that assess the efficacy of PDT against cancer and bacteria. Generally, it is used with hydrogels to improve the efficacy of PDT [30]. However, Law et al. conclude that additional research is required to evaluate the safety and cytotoxicity and improve the efficacy of PDT in various environments [31].

Discussion

In the context of this topic, there was an opportunity to appreciate both the clarity of the reviews and the proposed research articles, which include innovative techniques and treatment targets that will undoubtedly be further developed in the future. Based on the results of these studies, it can be concluded that PDT can generate growing optimism regarding the elimination of some diseases that have consistently plagued humanity.

Acknowledgments: We acknowledge the authors of the articles referred to in this editorial for their valuable contributions, and the referees for their rigorous review.

Conflicts of Interest: The author declares no conflicts of interest.

References

1. Grandi, V.; Corsi, A.; Pimpinelli, N.; Bacci, S. Cellular Mechanisms in Acute and Chronic Wounds after PDT Therapy: An Update. *Biomedicines* **2022**, *10*, 1624. [CrossRef] [PubMed]
2. Fernández-Guarino, M.; Bacci, S.; Pérez González, L.A.; Bermejo-Martínez, M.; Cecilia-Matilla, A.; Hernández-Bule, M.L. The Role of Physical Therapies in Wound Healing and Assisted Scarring. *Int. J. Mol. Sci.* **2023**, *24*, 7487. [CrossRef] [PubMed]
3. Nguyen, U.T.T.; Youn, E.; Le, T.A.N.; Ha, N.M.; Tran, S.H.; Lee, S.; Cha, J.W.; Park, J.S.; Kwon, H.C.; Kang, K. Photodynamic Treatment Increases the Lifespan and Oxidative Stress Resistance of Caenorhabditis elegans. *Free Radic. Biol. Med.* **2024**, *221*, 98. [CrossRef] [PubMed]
4. Muthubharathi, B.C.; Subalakshmi, P.K.; Mounish, B.S.C.; Rao, T.S.; Balamurugan, K. Impact of Low-dose UV-A in Caenorhabditis elegans during Candidate Bacterial Infections. *Photochem. Photobiol.* **2024**, *101*, 404–422. [CrossRef]
5. Wang, M.; Munch, I.C.; Hasler, P.W.; Prünte, C.; Larsen, M. Central serous chorioretinopathy. *Acta Ophthalmol.* **2008**, *86*, 126–145. [CrossRef]
6. Gawęcki, M.; Kiciński, K.; Kucharczuk, J.; Gołębiowska-Bogaj, M.; Grzybowski, A. Predictive Factors for Morphological and Functional Improvements in Long-Lasting Central Serous Chorioretinopathy Treated with Photodynamic Therapy. *Biomedicines* **2025**, *13*, 944. [CrossRef]
7. Aires-Fernandes, M.; Botelho Costa, R.; Rochetti do Amaral, S.; Mussagy, C.U.; Santos-Ebinuma, V.C.; Primo, F.L. Development of Biotechnological Photosensitizers for Photodynamic Therapy: Cancer Research and Treatment—From Benchtop to Clinical Practice. *Molecules* **2022**, *27*, 6848. [CrossRef]

8. Porolnik, W.; Karpinska, N.; Murias, M.; Piskorz, J.; Kucinska, M. Novel BODIPY Dyes with a *Meso*-Benzoxadiazole Substituent: Synthesis, Photophysical Studies, and Cytotoxic Activity Under Normoxic and Hypoxic Conditions. *Biomedicines* **2025**, *13*, 297. [CrossRef]
9. Bonta, M.; Daina, L.; Muţiu, G. The process of ageing reflected by histological changes in the skin. *Rom. J. Morphol. Embryol.* **2013**, *54*, 797–804.
10. Arnal-Forné, M.; Molina-García, T.; Ortega, M.; Marcos-Garcés, V.; Molina, P.; Ferrández-Izquierdo, A.; Sepulveda, P.; Bodí, V.; Ríos-Navarro, C.; Ruiz-Saurí, A. Changes in human skin composition due to intrinsic aging: A histologic and morphometric study. *Histochem. Cell Biol.* **2024**, *162*, 259–271. [CrossRef]
11. Harris, R.J.; Korolchuk, V.I. *Biochemistry and Cell Biology of Ageing, Part II Clinical Science*; Springer: Berlin, Germany, 2019.
12. Lee, H.; Hong, Y.; Kim, M. Structural and Functional Changes and Possible Molecular Mechanisms in Aged Skin. *Int. J. Mol. Sci.* **2021**, *22*, 12489. [CrossRef] [PubMed]
13. Notari, L.; Pieri, L.; Cialdai, F.; Fusco, I.; Risaliti, C.; Madeddu, F.; Bacci, S.; Zingoni, T.; Monici, M. Laser Emission at 675 nm: Molecular Counteraction of the Aging Process. *Biomedicines* **2024**, *12*, 2713. [CrossRef] [PubMed]
14. Fox, P.L.; Vasanji, A.; Ghosh, P.K.; Graham, L.M.; Eppell, S.J. Polarization of plasma membrane microviscosity during endothelial cell migration. *Dev. Cell* **2004**, *6*, 29–41. [CrossRef]
15. Michels, L.; Gorelova, V.; Harnvanichvech, Y.; Borst, J.W.; Albada, B.; Weijers, D.; Sprakel, J. Complete microviscosity maps of living plant cells and tissues with a toolbox of targeting mechanoprobes. *Proc. Natl. Acad. Sci. USA* **2020**, *117*, 18110–18118. [CrossRef]
16. Chambers, J.E.; Kubánková, M.; Huber, R.G.; López-Duarte, I.; Avezov, E.; Bond, P.J.; Marciniak, S.J.; Kuimova, M.K. An Optical Technique for Mapping Microviscosity Dynamics in Cellular Organelles. *ACS Nano.* **2018**, *12*, 4398–4407. [CrossRef]
17. Shimolina, L.E.; Khlynova, A.E.; Elagin, V.V.; Bureev, P.A.; Sherin, P.S.; Kuimova, M.K.; Shirmanova, M.V. Unraveling Microviscosity Changes Induced in Cancer Cells by Photodynamic Therapy with Targeted Genetically Encoded Photosensitizer. *Biomedicines* **2024**, *12*, 2550. [CrossRef]
18. Siegel, R.L.; Miller, K.D.; Fuchs, H.E.; Jemal, A. Cancer statistics, 2022. *CA Cancer J. Clin.* **2022**, *72*, 7–33. [CrossRef] [PubMed]
19. Wei, Y.; Niu, J.; Gu, L.; Hong, Z.; Bao, Z.; Qiu, L. Effect of Clinicopathological Characteristics on the Outcomes of Topical 5-Aminolevulinic Acid Photodynamic Therapy in Patients with Cervical High-Grade Squamous Intraepithelial Lesions (HSIL/CIN2): A Retrospective Cohort Study. *Biomedicines* **2024**, *12*, 2255. [CrossRef]
20. Ostrom, Q.T.; Bauchet, L.; Davis, F.G.; Deltour, I.; Fisher, J.L.; Langer, C.E.; Pekmezci, M.; Schwartzbaum, J.A.; Turner, M.C.; Walsh, K.M.; et al. The epidemiology of glioma in adults: A "state of the science" review. *Neuro-Oncology* **2014**, *16*, 896–913. [CrossRef]
21. Ostrom, Q.T.; Gittleman, H.; Truitt, G.; Boscia, A.; Krunko, C.; Barnholtz-Sloan, J.S. CBTRUS statistical report: Primary brain and central nervous system tumors diagnosed in united states in 2011–2015. *Neuro-Oncology* **2018**, *20*, iv1–iv86. [CrossRef]
22. Stupp, R.; Mason, W.P.; van den Bent, M.J.; Weller, M.; Fisher, B.; Taphoorn, M.J.; Belanger, K.; Brandes, A.A.; Marosi, C.; Bogdahn, U.; et al. European Organisation for Research and Treatment of Cancer Brain Tumor and Radiotherapy Groups, National Cancer Institute of Canada Clinical Trials Group Radiotherapy plus concomitant and adjuvant temozolomide for glioblastoma. *N. Engl. J. Med.* **2005**, *10*, 987–996. [CrossRef] [PubMed]
23. Saito, Y.; Fukami, S.; Nagai, K.; Ogawa, E.; Kuroda, M.; Kohno, M.; Akimoto, J. Cytocidal Effects of Interstitial Photodynamic Therapy Using Talaporfin Sodium and a Semiconductor Laser in a Rat Intracerebral Glioma Model. *Biomedicines* **2024**, *12*, 2141. [CrossRef] [PubMed]
24. Sung, H.; Ferlay, J.; Siegel, R.L.; Laversanne, M.; Soerjomataram, I.; Jemal, A.; Bray, F. Global Cancer Statistics 2020: GLOBOCAN Estimates of Incidence and Mortality Worldwide for 36 Cancers in 185 Countries. *CA A Cancer J. Clin.* **2021**, *12*, 209–249. [CrossRef]
25. Cancer Stat Facts: Pancreatic Cancer. Available online: https://seer.cancer.gov/statfacts/html/pancreas.html (accessed on 27 May 2025).
26. D'Antonio, D.L.; Marchetti, S.; Pignatelli, P.; Umme, S.; De Bellis, D.; Lanuti, P.; Piattelli, A.; Curia, M.C. Effect of 5-Aminolevulinic Acid (5-ALA) in "ALADENT" Gel Formulation and Photodynamic Therapy (PDT) against Human Oral and Pancreatic Cancers. *Biomedicines* **2024**, *12*, 1316. [CrossRef]
27. Bernier, S.P.; Surette, M.G. Concentration-dependent activity of antibiotics in natural environments. *Front. Microbiol.* **2013**, *4*, 20. [CrossRef]
28. Friedman, N.D.; Temkin, E.; Carmeli, Y. The negative impact of antibiotic resistance. *Clin. Microbiol. Infect.* **2016**, *22*, 416–422. [CrossRef] [PubMed]
29. Grin, M.; Suvorov, N.; Ostroverkhov, P.; Pogorilyy, V.; Kirin, N.; Popov, A.; Sazonova, A.; Filonenko, E. Advantages of combined photodynamic therapy in the treatment of oncological diseases. *Biophys. Rev.* **2022**, *14*, 941–963. [CrossRef]

30. Tian, B.; Liu, J. Resveratrol: A review of plant sources, synthesis, stability, modification and food application. *J. Sci. Food Agric.* **2020**, *100*, 1392–1404. [CrossRef]
31. Law, S.K.; Liu, C.W.C.; Tong, C.W.S.; Au, D.C.T. Potential of Resveratrol to Combine with Hydrogel for Photodynamic Therapy against Bacteria and Cancer—A Review. *Biomedicines* **2024**, *12*, 2095. [CrossRef]

biomedicines

MDPI

Article

Effect of 5-Aminolevulinic Acid (5-ALA) in "ALADENT" Gel Formulation and Photodynamic Therapy (PDT) against Human Oral and Pancreatic Cancers

Domenica Lucia D'Antonio [1,2], Simona Marchetti [1], Pamela Pignatelli [3], Samia Umme [1], Domenico De Bellis [4,5], Paola Lanuti [4,5], Adriano Piattelli [6,7] and Maria Cristina Curia [1,*]

1 Department of Medical, Oral and Biotechnological Sciences, "Gabriele d'Annunzio" University of Chieti-Pescara, 66100 Chieti, Italy; domenica.dantonio@unich.it (D.L.D.); simona.marchetti@phd.unich.it (S.M.); ummesamia95@gmail.com (S.U.)
2 Villa Serena Foundation for Research, Via Leonardo Petruzzi 42, 65013 Città Sant'Angelo, Italy
3 COMDINAV DUE, Nave Cavour, Italian Navy, Stazione Navale Mar Grande, Viale Jonio, 74122 Taranto, Italy; pamelapignatelli89p@gmail.com
4 Center for Advanced Studies and Technology (CAST), "Gabriele d'Annunzio" University of Chieti-Pescara, 66100 Chieti, Italy; domenico.debellis@unich.it (D.D.B.); p.lanuti@unich.it (P.L.)
5 Department of Medicine and Aging Sciences, "Gabriele d'Annunzio" University of Chieti-Pescara, 66100 Chieti, Italy
6 School of Dentistry, Saint Camillus International University of Health and Medical Sciences, Via di Sant'Alessandro 8, 00131 Rome, Italy; apiattelli51@gmail.com
7 Facultad de Medicina, UCAM Universidad Católica San Antonio de Murcia, 30107 Murcia, Spain
* Correspondence: mariacristina.curia@unich.it

Abstract: Oral squamous-cell and pancreatic carcinomas are aggressive cancers with a poor outcome. Photodynamic therapy (PDT) consists of the use of photosensitizer-induced cell and tissue damage that is activated by exposure to visible light. PDT selectively acts on cancer cells, which have an accumulation of photosensitizer superior to that of the normal surrounding tissues. 5-aminolevulinic acid (5-ALA) induces the production of protoporphyrin IX (PpIX), an endogenous photosensitizer activated in PDT. This study aimed to test the effect of a new gel containing 5% v/v 5-ALA (ALAD-PDT) on human oral CAL-27 and pancreatic CAPAN-2 cancer cell lines. The cell lines were incubated in low concentrations of ALAD-PDT (0.05%, 0.10%, 0.20%, 0.40%, 0.75%, 1.0%) for 4 h or 8 h, and then irradiated for 7 min with 630 nm RED light. The cytotoxic effects of ALAD-PDT were measured using the MTS assay. Apoptosis, cell cycle, and ROS assays were performed using flow cytometry. PpIX accumulation was measured using a spectrofluorometer after 10 min and 24 and 48 h of treatment. The viability was extremely reduced at all concentrations, at 4 h for CAPAN-2 and at 8 h for CAL-27. ALAD-PDT induced marked apoptosis rates in both oral and pancreatic cancer cells. Elevated ROS production and appreciable levels of PpIX were detected in both cell lines. The use of ALA-PDT as a topical or intralesional therapy would permit the use of very low doses to achieve effective results and minimize side effects. ALAD-PDT has the potential to play a significant role in complex oral and pancreatic anticancer therapies.

Keywords: 5-aminolevulinic acid; photodynamic therapy; pancreatic cancer; oral cancer; protoporphyrin; reactive oxygen species; CAL-27; CAPAN-2; apoptosis; cell cycle

1. Introduction

Pancreatic cancer (PC) and oral squamous-cell carcinoma (OSCC) are both lethal conditions with a poor outcome and an increasing incidence. In 2021, pancreatic and oral cancers accounted, ively, for 62,210 and 54,010 new cancer cases. Survival for both cancers is poor, with pancreatic cancer being the seventh-leading cause of cancer mortality worldwide, while oral cancer accounts for 1.8 percent of all cancer deaths [1,2]. Age-adjusted death rates of oral and pancreatic cancer have been rising, on average, by 0.4% and 0.5%, respectively, each year over 2010–2019 [3,4]. The average survival time for pancreatic cancer is low, partly because an early diagnosis is difficult. At present, surgical resection is the only potential cure for pancreatic cancer; however, only 20% of patients can proceed with surgical removal of the tumor [5] and rates of recurrence are high [6]. The oral cancer treatments available during recent decades have remained unchanged [7]. Among these, surgical resection, with or without adjuvant therapy, is a commonly used treatment. However, oral cancer is very aggressive and there is a high risk of developing second primary cancers. Unresectable oral cancers are treated with palliative systemic therapy and/or palliative radiotherapy [8] and with checkpoint inhibitors, acting on the tumor microenvironment [9].

Photodynamic therapy (PDT), whose general principle was first described in 1900, is an effective treatment for both precancerous and malignant conditions [10]. It consists of the use of a photosensitizer capable of inducing cell and tissue damage when activated by exposure to low-level visible light [11]. Several photosensitizers are activated by different total light doses, dose rates, and specific wavelengths of light radiation [11]. PDT selectively acts on cancer cells, which have an accumulation of photosensitizer superior to that of the normal surrounding tissues. In contrast to ionizing radiation, PDT is generally safer than other treatments for the surrounding normal tissues due to its cell selectivity and lack of cumulative toxicity, and it can be safely applied to previously irradiated tissues [12]. The introduction of 5-ALA in the Ppix metabolic pathway, regulated by the feedback control resulting from the presence of cellular concentrations of heme, leads to the intracellular accumulation of Ppix [13].

5-aminolevulinic acid (5-ALA) is metabolized to protoporphyrin IX, the final intermediate in the heme biosynthetic pathway, and an endogenous fluorescent photosensitizer. 5-ALA has been used to formulate a solution for treating potentially malignant lesions of the oral cavity by applying it topically through a saturated gauze placed against the injury or through a customized temporary dental prosthesis [14]. The application of 5-ALA in some lesions has proved difficult due to dilution by saliva and the difficulty of maintaining positioning. For this reason, it has been administered by intralesional injection or in a temperature-dependent gel-state formulation to optimize the bioavailability of the drug. A special sol–gel formula is marketed as ALADENT, which contains 5% of 5-ALA. It is composed of a mixture of poloxamers that ensure the stability of the active ingredient. This liquid thermolabile formulation transforms into gel at body temperature (>28 °C) [15,16]. The gel form allows better mucoadhesion, facilitating topical administration, greater hydrophilicity, relative solubility in water and lipids, and constant ionization [17]. This formulation showed a greater release of 5-ALA and, therefore, greater effectiveness in the accumulation of protoporphyrin in comparison with conventional cream and ointment vehicles, in the cure of dermatological and gastrointestinal disorders [18–21]. ALADENT is also used in the treatment of periodontitis and peri-implantitis thanks to its antibacterial activity against Gram-positive and -negative bacteria, such as Porphyromonas gingivalis, Enterococcus faecalis, Escherichia coli, Staphylococcus aureus, and Veillonella parvula, with or without irradiation. However, the strongest antibacterial effect was achieved with 25 min of 50% ALAD incubation followed by 5 min of a red LED [22].

PpIX, produced by 5-ALA metabolism, selectively kills neoplastic cells either directly, by stimulating the production of reactive oxygen species (ROS), or indirectly, by damaging tumor vascularity and activating immune responses against cancer cells [23]. The cytotoxic effect of PDT is due to cellular necrosis or apoptosis, depending on whether the 5-ALA localizes in lysosomes or cell membranes or penetrates mitochondria during light emission [24]. Cells with high rates of metabolic activity, such as cancer cells, inflammatory cells, and bacteria have lower amounts of ferrochelatase (FECH), an enzyme capable of metabolizing PpIX, so the production of photoactive porphyrins is more pronounced than in healthy cells [25]. Reduced FECH expression is present in various tumors such as kidney [26], bladder [27], and colorectal [28]. Low FECH activity of cancer cells causes accumulation of PpIX and other porphyrins in tumors, while normal or high FECH activity results in low accumulation of PpIX in tissues [29]. Thus, 5-ALA-based PDT can be a potential therapeutic strategy for the treatment of different tumors. In preclinical studies the anti-tumor efficacy of 5-ALA-mediated PDT has been observed in different cancer types such as skin cancer [30], esophageal cancer [31], colon cancer [28], human glioma [32], breast cancer [33], bladder cancer [34], and hepatocellular carcinoma [35].

A lot of illness, like head and neck [36], bladder [37], and Barrett's esophagus [38] cancers, as well as gynecological neoplasia [39] and actinic keratosis [40] are treated with 5-ALA-PDT. 5-ALA-PDT does not produce the same effects on all types of cancer cells; in fact, it can induce non-inflammatory cell death such as necrosis and apoptosis or inflammatory cell death. This therapy's use is greatly limited by the lack of knowledge of its mechanisms of action [41]. Other limitations to clinical application are the variability in 5-ALA uptake from cancer cells, the penetration of only 5 mm of 635 nm activating light, and poor oxygen recovery that limits the production of cytotoxic ROS [42]. As regards PC, in a previous paper, 5-ALA at higher doses followed by PDT prolonged animal survival [43].

ALAD intralesional or topical administration with in situ gel formation would allow the use of very low doses below the maximum dose allowed and used in humans, reducing systemic effects and resulting in effective 5-ALA concentrations. The use of ALAD-PDT could allow targeted anticancer therapy to be to carried out, with the possibility of repeating the administration over time with limited side effects. This new formulation of 5-ALA-mediated PDT was used for the first time on squamous oral and pancreatic cancer cells. The objective of this research was to examine how ALADENT affects oral and pancreatic cancer cells by measuring the level of PpIX, oxidative stress, apoptosis, and cell vitality. The aim was also to evaluate cell-cycle modulation after ALAD-PDT.

2. Materials and Methods

2.1. Chemicals

ALAD is a gel containing 5% v/v (4.6 M) of 5-ALA, commercialized as ALADENT by ALPHA Strumenti s.r.l. (Melzo, MI, Italy), and it is covered by European patent EP3727334 with the following description: "pharmaceutical preparation comprising a topically released active ingredient and a heat-sensitive carrier, method of obtaining same, and use of same in the treatment of skin and mucosal infections".

The full text of the patent is available at https://register.epo.org/espacenet/regviewer?AP=18836409&CY=EP&LG=en&DB=REG. It is a pharmaceutical preparation comprising an active ingredient 5-ALA, and a heat-sensitive carrier (thermogel) made of 19% poloxamer (PL) P407 and 4% poloxamer P188, with the balance being composed of water. Poloxamer hydrogels have a reversible thermo-responsive sol-to-gel transition (28 °C for ALADENT) which induces the formation of a solid gel from a solution, as a result of their self-assembling into micelles. ALAD remains stable when stored at 4° for up to a week, but after that, its effectiveness decreases. Poloxamer use is approved by the FDA and listed as pharmaceutical excipient in the United States and European Pharmacopoeias; the lack of toxic crosslinking agents renders it more likely than thermogels to show intrinsic biocompatibility as an injectable in situ hydrogel. The presence of a poloxamer mixture in the formulation facilitates the 5-ALA to rapidly enter into the mucous membranes of the target cells, acting as a transmucosal delivery.

2.2. Cell Lines and Culture

The investigations were performed on human tongue squamous carcinoma cell line CAL-27 and human pancreatic ductal adenocarcinoma cell line CAPAN-2. HGF-1 cells are a major constituent of the oral microenvironment, in particular of gingival connective tissue, which is the first tissue invaded by oral cancer. So they were used as a non-cancerous control cell line.

CAL-27 (ATCC-CRL-2095) and CAPAN-2 (CLS, Eppelheim, Germany) cancer cell lines were kindly provided by Dr. Maria Carmela Di Marcantonio and by Dr. Serena Veschi, respectively. HGF-1 cells (ATCC-CRL-2014) were purchased from LGC Standards S.r.l., Sesto San Giovanni, Milano, Italy.

CAL-27 and HGF-1 cells were cultured in Dulbecco's Modified Eagle's Medium (DMEM; EuroClone S.p.A., Pero, MI, Italy) and CAPAN-2 cells were cultured in a Roswell Park Memorial Institute medium (RPMI 1640, Corning, Glendale, AZ, USA). All the media were supplemented with 10% fetal bovine serum (FBS), 2 mM L-glutamine, and antibiotics (100 units/mL penicillin and 100 µg/mL streptomycin) (all from EuroClone).

Cultured cells were maintained in 5% CO_2 at 37 °C and 95% humidity and examined periodically under an inverted phase-contrast microscope. The cells grew as an adhering monolayer and the culture medium was changed once or twice a week. When cells reached 70–80% confluence, they were sub-cultured or seeded.

2.3. Light Source and Irradiation Parameters

An aluminium gallium arsenide (AlGaAs) power LED device (TL-01) characterized by 630 nm ± 10 nm full width half maximum (FHWM) nm wavelength was used as light source (Alphastrumenti, Melzo, MI, Italy). The hand-piece was constituted by 1 LED with 6 mm diameter at the exit and a surface irradiance of 380 mW/cm^2. During the experiments, the LED hand-piece was mounted perpendicularly to the wells at 0.5 mm of distance with a particular polystyrene box to maintain a constant distance from the light source, and also to obtain a uniform LED irradiation of all the samples. At these conditions, the exit irradiance surface was 380 mW/cm^2 and the total specific dose was 23 J/cm^2 for each minute of irradiation. The irradiation was performed under a laminar flow hood in the dark under aseptic conditions in all the experiments.

2.4. Cell Treatment

Twenty-four hours after the seeding, the CAL-27, CAPAN-2, and HGF-1 cells were incubated with increasing concentrations of ALAD: 0.05% v/v, 0.1% v/v, 0.2% v/v, 0.40% v/v, 0.75% v/v, and 1% v/v, corresponding, as reported in literature [44,45], to 5-ALA concentrations of 0.23 mM, 0.46 mM, 0.92 mM, 1.84 mM, 3.45 mM, and 4.6 mM, respectively, for different experimental times (2–8 h) in serum-free medium at 37 °C and 5% CO_2. Then, the cells were exposed to PDT using RED light (630 nm) (ALPHA Strumenti s.r.l) with an intensity of 380 mW/cm^2 for 7 min with a light dose of 23 J/cm^2. Subsequently, the serum-free medium was replaced with a medium containing 10% FBS. The control group consisted of untreated CAL-27 and CAPAN-2 (without ALAD) cells. The effects of ALAD-PDT on cell viability, apoptosis, the cell cycle, and ROS levels were assessed at selected concentrations and different time points, as indicated in the related paragraphs.

2.5. Cytotoxicity and Cell-Viability Assay (MTS Assay)

Cells in the logarithmic growth phase were seeded onto 96-well plates for 24 h, at different densities according to cell types: 8 × 10^3 cells/well for CAPAN-2, 10 × 10^3 cells/well for CAL-27, and 6 × 10^3 cells/well for HGF-1. Then, cell lines were treated with ALAD-PDT, as reported above.

After 24 h, MTS solution (10 µL) (Promega's CellTiter 96® AQueous Non-Radioactive Cell Proliferation Assay, Madison, WI, USA) was added to each well, followed by incubation at 37 °C, 5% CO_2 for 1 h. All conditions were performed as quintuplicate. The absorbance was determined at 490 nm using a microplate reader (Synergy H1 Hybrid BioTek Instruments). Cell viability was reported as a percentage as compared with the untreated cells (CTRL), recognized as 100%.

2.6. Flow Cytometry Analysis

2.6.1. Detection of Apoptosis

To discriminate apoptotic cells, propidium iodide (PI) (BD Biosciences, Milano, Italy), which is able to identify damaged membranes, and Annexin V-FITC (fluorescein-isothiocyanate) (BD Biosciences), which allows the detection of phosphatidylserine expression on cell membranes, were used to stain 3×10^5 cells (15 min at RT, 25 °C in the dark) [46]. The use of Annexin V to stain the phosphatidylserine flipped on the external layer of the plasma membrane provides a much more comprehensive picture of the apoptotic process than the identification of the sub-G1 peak [47,48]. Therefore, the AnnexinV/PI analysis was applied to study the ALAD-PDT-induced apoptosis.

Samples were then analyzed by flow cytometry using a FACSVerse cytometer (BD Biosciences). Finally, data were analyzed using FlowJo v10.10.0 software (BD Biosciences, Milano, Italy).

2.6.2. Cell-Cycle Analysis

An amount of 5×10^5 cells per sample were harvested, fixed in 70% cold ethanol and stored at 4 °C. Samples were then resuspended in 50 µg/mL PI (Sigma, St. Louis, MO, USA) and 120 µg/mL RNAse (Sigma, St. Louis, MO, USA), as previously described [49]. Samples were acquired using a FACScanto II flow cytometer (BD Biosciences). Data were analyzed with FlowJo software v10.10.0 (BD Biosciences).

2.6.3. Reactive Oxygen Species (ROS) Levels

Reactive oxygen species production was carried out by staining the cells (5×10^5 cells/sample) with 5 µM of 5-(and-6)-chloromethyl-2′,7′-dichlorodihydrofluorescein diacetate, acetyl ester (CM-H2DCFDA Molecular Probes, Invitrogen, Life-Sciences-Division, Milan, Italy) for 1 h at 37 °C. When ROS are produced, CM-H2DCFDA is oxidized and an increase in green fluorescence is detectable by flow cytometry. Samples were analyzed by using a FACSVerse cytometer (BD Biosciences, San Jose, CA, USA). Data were analyzed using FlowJo v10.10.0 (BD Biosciences).

2.7. PpIX Fluorescence Measurements

PpIX intracellular content was determined after ALAD-PDT treatment (4 h for CAPAN-2 and 8 h for CAL-27) at different time points, namely T1 (10 min after the treatment), T2 (24 h after the treatment), and T3 (48 h after the treatment).

An amount of 8×10^3 cells/well for CAPAN-2 and 10×10^3 cells/well for CAL-27 were plated onto 96-well plates and subjected to the ALAD-PDT treatment. Then, cells were treated with a solution of perchloric acid (HClO4) in methanol, as previously described [50], and PpIX fluorescence was measured using a microplate spectrofluorometer (Synergy H1 Hybrid BioTek Instruments, Winooski, VT, USA) at λex/em 405/608.

2.8. Statistical Analysis

Data are expressed as the means ± standard deviation (SD). Statistical analyses were performed using the *t*-test to compare treated to control samples. Differences between groups were assessed with one-way analysis of variance (ANOVA). A p-value ≤ 0.05 was considered as significant. Statistical analyses and descriptive statistics were carried out using GraphPad Prism version 9.0 (GraphPad Software Inc., La Jolla, CA, USA).

3. Results

3.1. Cytotoxicity

The CAL-27 and CAPAN-2 human cancer cell lines were incubated at low concentrations of ALAD (0.05%, 0.10%, 0.20%, 0.40%, 0.75%, 1%) at different times and then exposed to 630 nm RED light irradiation. At first, ALAD-PDT-treated cells were compared with ALAD-treated cells without LED irradiation (Figure S1). Results showed marked differences with the use of light and the study continued only using the ALAD-PDT treatment. Furthermore, the results confirmed literature data on the use of PDT.

After ALAD-PDT, for CAPAN-2 cells, the optimum inhibition efficiencies were obtained at 4 h, and for CAL-27 cells, at 8 h. Very low concentrations of ALD-PDT blocked cell growth and caused apoptosis in both cell lines (Figure 1). The differences were in the extent of cytotoxicity and in the treatment time—4 h for CAPAN-2 cells and 8 h for CAL-27 cells. At the lowest concentration of 0.05% of ALADENT, there was a notable decrease in vitality in both cell lines. For this reason, subsequent analyzes were performed on the two lower concentrations, 0.05% and 0.1%, which corresponded to the "minimum effective dose" (MED).

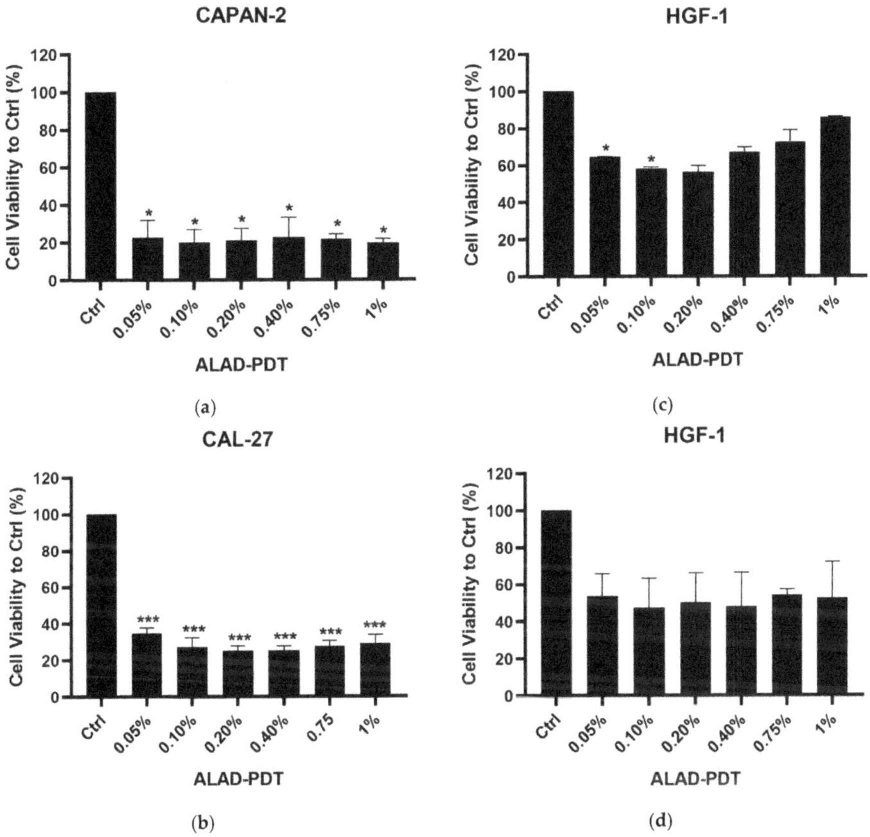

Figure 1. MTS assay after ALAD-PDT treatment. Cell viability of CAPAN-2 cells after 4 h (**a**) and of CAL-27 cells after 8 h (**b**) of treatment. Cell viability of HGF-1 cells after 4 h (**c**) and 8 h (**d**) of treatment. Data shown are means ± standard deviation (SD) of two independent experiments with quintuplicate determinations. * Statistically significant differences as compared with control (* $p < 0.05$; *** $p < 0.001$).

For CAL-27, the viability was reduced by approximately 70% at all concentrations after 8 h (Figure 1). CAPAN-2 cells showed a higher sensitivity to ALAD-PDT than CAL-27 cells and already, after 4 h incubation, an almost total reduction (80%) in viability was visible (Figure 1) at all concentrations.

The MTS assay was also performed on human gingival fibroblast (HGF-1) cells under the same conditions to evaluate the selective toxicity of ALAD-PDT on normal cells (Figure 1). The results showed a percentage of cell viability of not less than approximately 50% for both experimental times (4 and 8 h) after treatment.

3.2. *Apoptosis Rates and Cell-Cycle Analysis*

The influence of ALAD-PDT on the induction of apoptotic cell death was studied by evaluating exposure to phosphatidylserine through Annexin-V staining of CAL-27 and CAPAN-2 cells. For CAPAN-2, given the high cell mortality after 4 h of treatment, the treatment time was reduced to 2 h, maintaining low ALAD treatment concentrations. Figure 2 shows the gating strategy used to identify Annexin-V-positive cells. Briefly, debris was excluded on an FSC-A/SSC-A dot plot and the four sub-populations identified by PI and Annexin V staining were gated. In Figure 3, the fold-change values of the percentage of total Annexin-V-positive cells, calculated with respect to untreated samples, were reported as bars both for 0.05% and 0.1% ALAD-PDT treatments (CAPAN-2: p-value = 0.3284, 95% confidence interval −35.9669 to 84.0339; CAL-27: p-value = 0.0186, 95% confidence interval 19.4935 to 76.9067). As shown, the induction of apoptosis was increased in ALAD-PDT-treated CAL-27 and CAPAN-2 cells with respect to untreated samples. This effect was dose-dependent with a significant increase in apoptosis between 0.05% and 0.1% ALAD-PDT treatments in CAL-27 cells. Such an increase was also detected in CAPAN-2 cells but without reaching statistical significance.

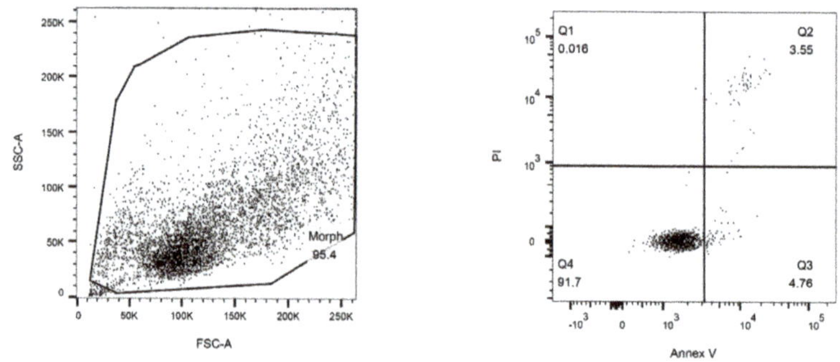

(a) (b)

Figure 2. Gating strategy for apoptosis detection: (**a**) All events were gated (morph region) on a Forward Scatter Area (FSC-A)/Side Scatter Area (SSC-A) dot plot. (**b**) Cells were analyzed on a dot plot of Annexin V (Annex V)/propidium iodide (PI). The identified regions are as follows: Q1 = necrotic cells (Annex V−/PI+), Q2 = late apoptotic cells (Annex V+/PI+), Q3 = early apoptotic cells (Annex V+/PI−), and Q4 = live cells (Annex V−/PI−). The represented gating strategy refers to the CAPAN-2 cell line analyzed under logarithmic growth conditions. The same gating strategy was used to analyze all samples.

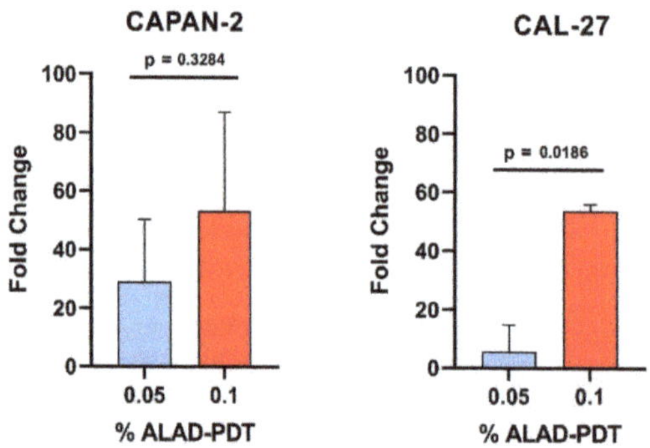

(a) (b)

Figure 3. Flow cytometry analysis of apoptosis induced by ALAD-PDT treatment. Bars represent the percentage of fold increase in cells staining positive to Annexin V after cell treatment with two different concentrations, 0.05% (blue bars) and 0.1% (red bars), of ALAD (ALAD-PDT), with respect to the untreated sample set at 1 (CTRL) and, therefore, not represented, both for CAPAN-2 (**a**) and CAL-27 (**b**) cell lines. Percentage of fold-change values were calculated as the ratio between the percentage of Annexin-V-positive cells induced by each treatment (0.05% or 0.1%) and the percentage of Annexin-V-positive cells detected in the respective untreated sample (CTRL). Data are presented as the means ± SD of duplicate experiments. $p < 0.05$ indicates statistical significance.

Cell-cycle analysis was also carried out (Figure 4). As shown, the ALAD-PD treatment did not impact the cell cycle at either at 0.05% or 0.1%.

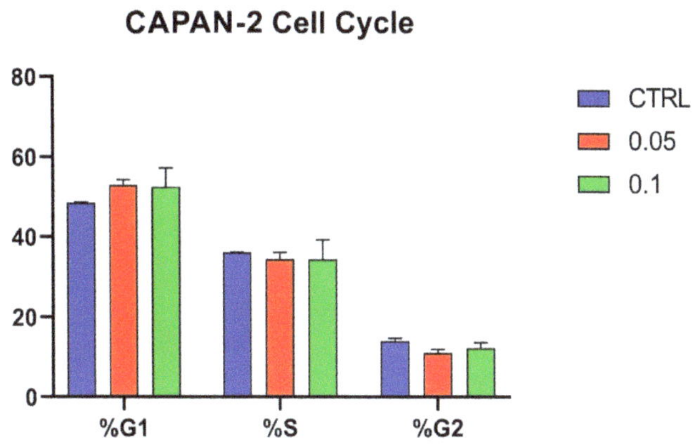

(**a**)

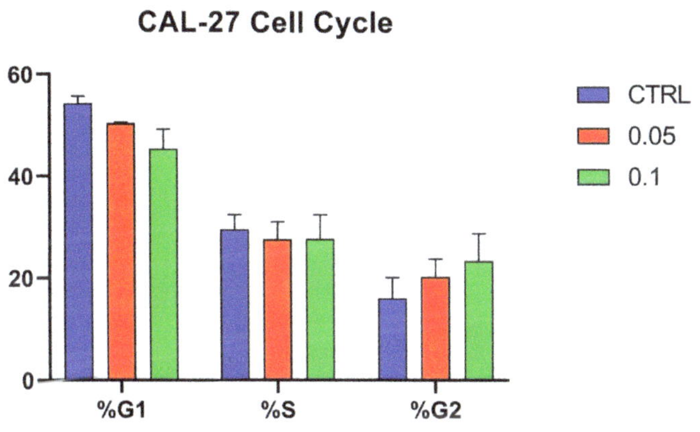

(**b**)

Figure 4. Effect of ALAD-PDT on CAPAN-2 (**a**) and CAL-27 (**b**) cell cycle. Bars represent the percentages (y-axis) of cell distribution throughout the cell-cycle progression (G1, S, and G2 phases) in controls (CTRL, blue bars) and after cell treatment with two different concentrations, 0.05% (red bars) and 0.1% (green bars), of ALAD (x-axis). Data were analyzed using FlowJo v10.10.0 software and are presented as the means ± SD of duplicate experiments.

3.3. ROS-Level Variation

ALAD is enzymatically converted into PpIX, that is photoactivated by light and produces ROS, inducing cell death. The fluorescence measured by the DCFH2-DA assay was used to study the ROS production. The ratio between the DCFH2-DA fluorescence of 0.05% or 0.1% ALAD-PDT-treated cells and that of untreated samples, is reported as bars in Figure 5. As shown, the ROS production was increased in ALAD-PDT-treated CAL-27 and CAPAN-2 cells with respect to untreated samples. This effect was dose-dependent with a significant increase in ROS production between 0.05% and 0.1% ALAD-PDT treatments in CAL-27 cells (CAPAN-2: *p*-value 0.0676, 95% confidence interval −0.271360 to 3.29536; CAL-27: *p*-value 0.0034, 95% confidence interval 0.762759 to 1.27399). Such an increase was also detected in CAPAN-2 cells but without reaching statistical significance (Figure 5).

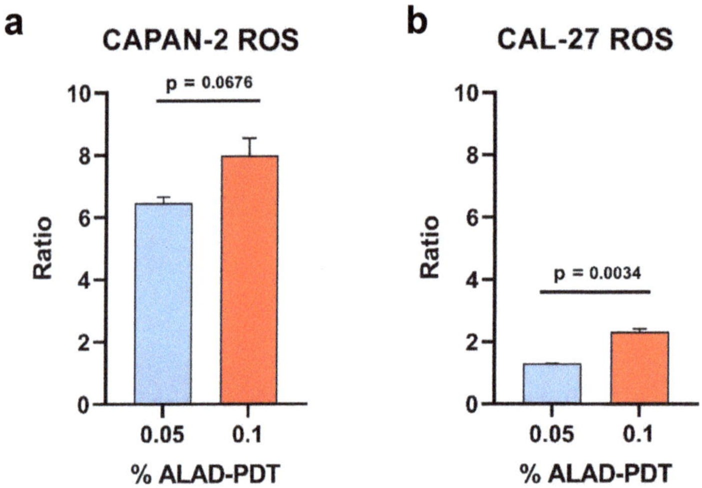

Figure 5. The effect of ALAD-PDT on ROS production by flow cytometry. Bars represent the ratio of ROS production after cell treatment with two different concentrations of ALAD (ALAD-PDT), 0.05% (blue bars) and 0.1% (red bars), and the untreated samples set at 1 (CTRL) and, therefore, not represented, both for CAPAN-2 (**a**) and CAL-27 (**b**) cell lines. Ratio values were calculated as the ratio between the mean fluorescence intensity (MFI) of each treatment (0.05% or 0.1%) and the MFI of the untreated sample (CTRL). Data are presented as the means ± SD of duplicate experiments. $p < 0.05$ indicates statistical significance.

These results demonstrate that the production of reactive oxygen species was stimulated by photoactivation.

3.4. Generation of Intracellular PpIX

PpIX accumulation analysis was also carried out. PpIX fluorescence was measured in CAPAN-2 and CAL-27 cell lines showing an increase in PpIX that occurred within 10 min (T1) after the treatment time (4 h for CAPAN-2 and 8 h for CAL-27) with ALAD-PDT. The PpIX content was maintained over time (T2, 24 h; T3, 48 h), with slight increases and decreases based on the concentrations of ALAD (0.05% and 0.1%) and on cell type (Figure S2). Data showed statistically significant differences as compared with the control.

4. Discussion

OSCC and PC are characterized by an increase in incidence worldwide and they are associated with a high mortality rate. This demands the development of new strategies to treat these deadly malignancies. PDT is an alternative therapy that relies on the activation of a light-sensitive photosensitizer using visible light in the presence of molecular oxygen, causing ROS production [51]. The process then leads to irreversible cell damage and localized cell death [52,53]. Numerous advantages, including minimal toxicity to normal tissues, negligible systemic effects, protection of organ function, and high efficacy, make PDT a medical technique for the treatment of cancer patients [54]. The photosensitizer holds the major role in anti-tumor activity [44]. In cells, 5-ALA is converted to protoporphyrin IX (PpIX), a photosensitizer and direct precursor of hemoglobin in the heme synthesis pathway, which accumulates in mitochondria, preferentially in the tumor. This occurs as a result of the reduced activity of ferrochelatase and the lower availability of iron [55]. Considering that the absorbance range of 5-ALA is between 420 and 760, in vitro, we chose a wavelength of 630 nm to observe a high PDT effect. 5-ALA has an anti-inflammatory and immunoregulatory action reducing the expression of Toll-like receptor (TLR) 2 and 4 [56].

PDT is usually defined as a powerful inducer of tumor-cell apoptosis [33]. The rate of apoptosis increased after ALAD-PDT, a sign that its administration causes an arrest in the proliferation of both cell lines but at different times. In fact, for CAPAN-2 cells, the incubation time was reduced due to the high mortality of cells post ALAD-PDT. Another important role in apoptotic cell death is the formation of mitochondrial ROS after ALAD-PDT. It is well known that mitochondrial ROS formation plays a critical role in PDT-induced apoptosis. Mitochondrial inner-membrane permeabilization leads to mitochondrial swelling, cytochrome c release to the cytosol, and activation of the caspase cascade [57]. On the other hand, when singlet oxygen and lipid peroxide induced by ALAD-PDT destroy the plasma membrane, the cell becomes necrotic. It has been reported that the cytosol localization of PpIX caused cell necrosis, while photosensitizer localizing in mitochondria and lysosomes led to more pronounced apoptosis [58].

Mitochondrial ROS production can be of enormous importance for cell proliferation and apoptosis [34,35]. Recently, mitochondria-targeting therapeutic strategies have been considered as a novel approach for cancer treatment. As reported in literature, ALA-induced PpIX is initially localized in the mitochondria [36]. Excessive mitochondrial ROS production may lead to cell death induced by oxidative stress; however, these responses differ in different cell types [37]. Our results demonstrated that ALAD-PDT influenced ROS production in CAPAN-2 cells more than in CAL-27 cells. Thus, we assume that induced mitochondrial ROS production probably provides a clue to the differential susceptibility of CAL-27 and CAPAN cells to ALAD-PDT.

ROS levels were increased in both treated cell lines with respect to untreated samples. The increase was marked in CAPAN-2 cells at both ALAD concentrations. On the other hand, CAL-27 cells showed a significant increase in both apoptosis and ROS production between 0.05% and 0.1% ALAD concentrations, indicating 0.1% as the most effective treatment. These observations confirm literature data on esophageal cancer cells during PDT in mice [59].

It has been shown that lower 5-ALA concentrations (1–4 mM) allowed an inhibitory effect after 2–16 h incubation in oral and pancreatic cancer cell lines [44,45].The formulation of 5% 5-ALA with poloxamer mixture facilitates uptake from target cells, reducing the minimum inhibitory concentration of 5-ALA compared with previous concentrations used for PDT. Using a drug at low concentration reduces possible side effects.

This is the first in vitro study investigating the differences in sensitivity to 5-ALA-PDT between OSCC and PC. Interestingly, CAL-27 and CAPAN-2 cells responded similarly to ALAD-PDT treatment even if ALAD-PDT showed different effects on the two cell lines. In fact, CAL-27 cells were more resistant to treatment, reaching a relevant cytotoxic effect after only 8 h of treatment. These findings are consistent with the pathobiological behavior of OSCC cells, which have a higher metabolic rate and proliferate faster [30,31]. This is probably also due to the greater ability of the cells of the oral cavity to resist mechanical and thermal stress, due to their anatomical location. Confirming this, in the study by Rosin et al., the *SCC9* cell line, isolated from the tongue of a patient with squamous-cell carcinoma, survived after 5-ALA-PDT administration by reducing PpIX synthesis and initiating signaling pathways related to cell proliferation and apoptosis [60].

On the contrary, the CAPAN-2 cell line showed a greater response to the treatment after 2 h, the first incubation time tested. Little is known about the in vitro effect of 5-ALA-PDT on human PC cell lines; Khaled et al. assessed the anti-cancer effects of 5-ALA-PDT only when it was combined with oncolytic viral therapy [45].

Preclinical and clinical studies of 5-ALA-PDT against oral and pancreatic cancer showed positive and promising results. In the study by Zhu et al. of oral squamous carcinoma, 5-ALA-mediated PDT both in vitro and in vivo triggered the generation of intracellular ROS with significant cytotoxicity and apoptosis-induction effects, thus inhibiting tumor growth [61]. Furthermore, in the study by Wang et al., topical 5-ALA-PDT proved to be a safe treatment in OSCC patients with locally advanced sites, in addition to platinum-based induction chemotherapy (ICT) [62]. Despite the short observation period and small sample size, the results suggest prospective studies to evaluate the efficacy and safety of topical 5-ALA-PDT as an adjuvant to ICT followed by surgery.

As regards PC treatment, in the study by Regula et al., 5-ALA was administered systemically (intravenously or orally) to animals, with higher doses prolonging animal survival [47]. In another study, the prodrug 5-ALA was selectively activated by endogenous Cathepsin E within PC cells in vitro and in vivo [63]. This new combined approach reduced the viability of pancreatic cancer cells, with few side effects, while sparing normal cells.

ALADENT intralesional or topical administration with in situ gel formation would allow the use of very low doses, below the maximum dose allowed and used in humans, reducing systemic effects and resulting in effective 5-ALA concentrations.

Limitations and Future Research Directions

This was a preliminary and in vitro study. Further studies to identify the molecular signaling pathways involved in the cell death mechanisms in 5-ALAD-PDT-exposed cancer are required. Only a few functions and possible mechanisms of 5-ALAD-PDT have been elucidated in this research. A better understanding of the pathophysiological characteristics of oral and PC cells is essential for exploring the mechanism of 5-ALA-PDT. Additionally, further studies should be devoted to determining how other related molecules and signaling pathways interact with each other in the process.

Therefore, additional investigations are needed to translate 5-ALAD-PDT findings on alternative therapeutic strategies. Finally, future studies should focus on in vivo functions and mechanisms, which may provide better evidence to facilitate our understanding of the potential clinical implications of 5-ALAD-PDT, considering that various cell lines respond in different ways to this treatment.

5. Conclusions

Following the results obtained from experiments, we can confirm that ALAD-PDT markedly inhibited the viability and proliferation of PC and OSCC cell lines through mechanisms involving accumulation of PpIX that, under irradiation, can induce profound ROS production and phototoxicity. When cells were treated with low concentrations of ALAD-PDT, the intracellular PpIX level increased producing a large amount ROS. The results of the study indicate that administered 5-ALA could be successfully interconverted to PpIX in cells, inducing focal cancer-cell death. The absorption of the active substance would be facilitated by this new gel formulation, which would make treatment of OSCC-affected regions, such as the tongue and lingual floor, more convenient and non-invasive. Also, ALAD intralesional administration would allow its application in pancreatic-cancer treatment. By performing localized treatment of the pathological area in the early stages, it would be possible to avoid more invasive surgical and chemotherapy treatments. The initial treatment of oral cancers with ALAD-PDT would reduce the extent and contain the pathology as adjuvant therapy.

PDT and ALAD together slowed the growth of PC and OSCC cells, presenting itself as a promising treatment strategy for oral and pancreatic cancer. Notably, the time responses of the CAL-27 and CAPAN-2 cells were different, probably due to differences in cell type and tumor location. Also, for pancreatic cancer, there may be promising prospects for conservative therapies following studies that can provide clarification. Indeed, several in vitro and in vivo studies need to be performed to better understand the potential of ALAD-PDT as a targeted therapy for cancer. In our research, we have only explored and addressed a few of the possible mechanisms of ALAD-PDT.

These findings indicate that ALAD-PDT may be the breakthrough in anti-cancer adjuvant treatment.

Supplementary Materials: The following supporting information can be downloaded at: https://www.mdpi.com/article/10.3390/biomedicines12061316/s1, Figure S1: MTS assay after ALAD treatment with and without PDT; Figure S2: Levels of intracellular PpIX.

Author Contributions: Methodology, D.L.D., S.M., P.P., S.U., D.D.B. and P.L.; supervision of the project, A.P.; conceptualization and supervision of the project, M.C.C. All authors have read and agreed to the published version of the manuscript.

Funding: This research received no external funding.

Institutional Review Board Statement: Not applicable.

Informed Consent Statement: Not applicable.

Data Availability Statement: Data is contained within the article or Supplementary Materials.

Acknowledgments: ALADENT gel was supported by ALPHA Strumenti s.r.l. (Melzo, MI), Italy.

Conflicts of Interest: The authors declare no conflicts of interest. Pending patent application n. 102022000015318.

Abbreviations

PC	Pancreatic cancer
OSCC	Oral squamous-cell carcinoma
5-ALA	5-aminolevulinic acid
PpIX	Protoporphyrin IX
PDT	Photodynamic therapy
ROS	Reactive oxygen species
ALAD	ALADENT
HGF-1	Human gingival fibroblast
FECH	Ferrochelatase
HClO4	Perchloric acid
Control	CTRL
MED	Minimum effective dose
PL	Poloxamer
SD	Standard deviation
ANOVA	Analysis of variance
PI	Propidium iodide
FITC	Fluorescein-isothiocyanate
DMEM	Dulbecco's Modified Eagle's Medium
FBS	Fetal bovine serum
TLR	Toll-like receptor
AlGaAs	Aluminium gallium arsenide
FHWM	Full width half maximum
CM-H2DCFDA	5-(and-6)-chloromethyl-2′,7′-dichlorodihydrofluorescein diacetate, acetyl ester

References

1. International Agency for Research on Cancer (IARC). Globocan 2020. Available online: https://gco.iarc.fr/today/data/fact-sheets/cancers/13-Pancreas-fact-sheet.pdf (accessed on 12 May 2024).
2. Cancer Stat Facts: Pancreatic Cancer. Available online: https://seer.cancer.gov/statfacts/html/pancreas.html (accessed on 12 May 2024).

3. Cancer Stat Facts: Oral Cavity and Pharynx Cancer. Available online: https://seer.cancer.gov/statfacts/html/oralcav.html (accessed on 12 May 2024).
4. Bray, F.; Ferlay, J.; Soerjomataram, I.; Siegel, R.L.; Torre, L.A.; Jemal, A. Global cancer statistics 2018: GLOBOCAN estimates of incidence and mortality worldwide for 36 cancers in 185 countries. *CA Cancer J. Clin.* **2018**, *68*, 394–424. [CrossRef]
5. McGuigan, A.; Kelly, P.; Turkington, R.C.; Jones, C.; Coleman, H.G.; McCain, R.S. Pancreatic cancer: A review of clinical diagnosis, epidemiology, treatment and outcomes. *World J. Gastroenterol.* **2018**, *24*, 4846–4861. [CrossRef]
6. Baek, B.; Lee, H. Prediction of survival and recurrence in patients with pancreatic cancer by integrating multi-omics data. *Sci. Rep.* **2020**, *10*, 18951. [CrossRef] [PubMed]
7. Graboyes, E.M.; Kompelli, A.R.; Neskey, D.M.; Brennan, E.; Nguyen, S.; Sterba, K.R.; Warren, G.W.; Hughes-Halbert, C.; Nussenbaum, B.; Day, T.A. Association of Treatment Delays with Survival for Patients with Head and Neck Cancer: A Systematic Review. *JAMA Otolaryngol. Head Neck Surg.* **2019**, *145*, 166–177. [CrossRef] [PubMed]
8. Patil, V.M.; Noronha, V.; Joshi, A.; Abhyankar, A.; Menon, N.; Dhumal, S.; Prabhash, K. Beyond conventional chemotherapy, targeted therapy and immunotherapy in squamous cell cancer of the oral cavity. *Oral. Oncol.* **2020**, *105*, 104673. [CrossRef]
9. Parmar, K.; Mohamed, A.; Vaish, E.; Thawani, R.; Cetnar, J.; Thein, K.Z. Immunotherapy in head and neck squamous cell carcinoma: An updated review. *Cancer Treat. Res. Commun.* **2022**, *33*, 100649. [CrossRef]
10. Li, Y.; Wang, B.; Zheng, S.; He, Y. Photodynamic therapy in the treatment of oral leukoplakia: A systematic review. *Photodiagnosis Photodyn. Ther.* **2019**, *25*, 17–22. [CrossRef]
11. Correia, J.H.; Rodrigues, J.A.; Pimenta, S.; Dong, T.; Yang, Z. Photodynamic Therapy Review: Principles, Photosensitizers, Applications, and Future Directions. *Pharmaceutics* **2021**, *13*, 1332. [CrossRef] [PubMed]
12. Wang, X.; Luo, D.; Basilion, J.P. Photodynamic Therapy: Targeting Cancer Biomarkers for the Treatment of Cancers. *Cancers* **2021**, *13*, 2992. [CrossRef]
13. Markina, Y.V.; Markin, A.M.; Kirichenko, T.V.; Tolstik, T.V.; Cherednichenko, V.R.; Kiseleva, D.G.; Orekhov, A.N. Effect of 5-aminolevulinic Acid on Mitochondrial Activity. *Front. Biosci.* **2023**, *15*, 17. [CrossRef]
14. Maloth, K.N.; Velpula, N.; Kodangal, S.; Sangmesh, M.; Vellamchetla, K.; Ugrappa, S.; Meka, N. Photodynamic Therapy—A Non-invasive Treatment Modality for Precancerous Lesions. *J. Lasers Med. Sci.* **2016**, *7*, 30–36. [CrossRef] [PubMed]
15. Zarrintaj, P.; Ramsey, J.D.; Samadi, A.; Atoufi, Z.; Yazdi, M.K.; Ganjali, M.R.; Amirabad, L.M.; Zangene, E.; Farokhi, M.; Formela, K.; et al. Poloxamer: A versatile tri-block copolymer for biomedical applications. *Acta Biomater.* **2020**, *110*, 37–67. [CrossRef] [PubMed]
16. Luo, L.; Zhang, Q.; Luo, Y.; He, Z.; Tian, X.; Battaglia, G. Thermosensitive nanocomposite gel for intra-tumoral two-photon photodynamic therapy. *J. Control. Release* **2019**, *298*, 99–109. [CrossRef] [PubMed]
17. Del Duca, E.; Manfredini, M.; Petrini, N.; Farnetani, F.; Chester, J.; Bennardo, L.; Schipani, G.; Tamburi, F.; Sannino, M.; Cannarozzo, G.; et al. Daylight photodynamic therapy with 5-aminolevulinic acid 5% gel for the treatment of mild-to-moderate inflammatory acne. *Ital. J. Dermatol. Venereol.* **2021**, *156*, 46–50. [CrossRef] [PubMed]
18. Collaud, S.; Peng, Q.; Gurny, R.; Lange, N. Thermosetting gel for the delivery of 5-aminolevulinic acid esters to the cervix. *J. Pharm. Sci.* **2008**, *97*, 2680–2690. [CrossRef] [PubMed]
19. Serini, S.M.; Cannizzaro, M.V.; Dattola, A.; Garofalo, V.; Del Duca, E.; Ventura, A.; Milani, M.; Campione, E.; Bianchi, L. The efficacy and tolerability of 5-aminolevulinic acid 5% thermosetting gel photodynamic therapy (PDT) in the treatment of mild-to-moderate acne vulgaris. A two-center, prospective assessor-blinded, proof-of-concept study. *J. Cosmet. Dermatol.* **2019**, *18*, 156–162. [CrossRef] [PubMed]
20. Scarabello, A.; Pulvirenti, C.; Adebanjo, G.A.R.; Parisella, F.R.; Chello, C.; Tammaro, A. Photodynamic therapy with 5 aminolaevulinic acid: A promising therapeutic option for the treatment of Hailey-Hailey disease. *Photodiagnosis Photodyn. Ther.* **2022**, *38*, 102794. [CrossRef] [PubMed]
21. Bourre, L.; Thibaut, S.; Briffaud, A.; Lajat, Y.; Patrice, T. Potential efficacy of a delta 5-aminolevulinic acid thermosetting gel formulation for use in photodynamic therapy of lesions of the gastrointestinal tract. *Pharmacol. Res.* **2002**, *45*, 159–165. [CrossRef] [PubMed]
22. Radunović, M.; Petrini, M.; Vlajic, T.; Iezzi, G.; Di Lodovico, S.; Piattelli, A.; D'Ercole, S. Effects of a novel gel containing 5-aminolevulinic acid and red LED against bacteria involved in peri-implantitis and other oral infections. *J. Photochem. Photobiol. B* **2020**, *205*, 111826. [CrossRef]
23. Pignatelli, P.; Umme, S.; D'Antonio, D.L.; Piattelli, A.; Curia, M.C. Reactive Oxygen Species Produced by 5-Aminolevulinic Acid Photodynamic Therapy in the Treatment of Cancer. *Int. J. Mol. Sci.* **2023**, *24*, 8964. [CrossRef]
24. Wang, Y.Y.; Chen, Y.K.; Hu, C.S.; Xiao, L.Y.; Huang, W.L.; Chi, T.C.; Cheng, K.H.; Wang, Y.M.; Yuan, S.F. MAL-PDT inhibits oral precancerous cells and lesions via autophagic cell death. *Oral. Dis.* **2019**, *25*, 758–771. [CrossRef] [PubMed]
25. Harada, Y.; Murayama, Y.; Takamatsu, T.; Otsuji, E.; Tanaka, H. 5-Aminolevulinic Acid-Induced Protoporphyrin IX Fluorescence Imaging for Tumor Detection: Recent Advances and Challenges. *Int. J. Mol. Sci.* **2022**, *23*, 6478. [CrossRef] [PubMed]
26. Zhong, G.; Li, Q.; Luo, Y.; Liu, Y.; Liu, D.; Li, B.; Wang, T. FECH Expression Correlates with the Prognosis and Tumor Immune Microenvironment in Clear Cell Renal Cell Carcinoma. *J. Oncol.* **2022**, *2022*, 8943643. [CrossRef] [PubMed]
27. Nakai, Y.; Tatsumi, Y.; Hori, S.; Morizawa, Y.; Iida, K.; Onishi, K.; Miyake, M.; Oda, Y.; Owari, T.; Fujii, T.; et al. 5-Aminolevurinic acid inhibits the proliferation of bladder cancer cells by activating heme synthesis. *Oncol. Rep.* **2022**, *48*, 186. [CrossRef] [PubMed]

28. Yamada, K.; Murayama, Y.; Kamada, Y.; Arita, T.; Kosuga, T.; Konishi, H.; Morimura, R.; Shiozaki, A.; Kuriu, Y.; Ikoma, H.; et al. Radiosensitizing effect of 5-aminolevulinic acid in colorectal cancer. *Oncol. Lett.* **2019**, *17*, 5132–5138. [CrossRef] [PubMed]
29. Palasuberniam, P.; Kraus, D.; Mansi, M.; Braun, A.; Howley, R.; Myers, K.A.; Chen, B. Ferrochelatase Deficiency Abrogated the Enhancement of Aminolevulinic Acid-mediated Protoporphyrin IX by Iron Chelator Deferoxamine. *Photochem. Photobiol.* **2019**, *95*, 1052–1059. [CrossRef] [PubMed]
30. Sun, J.; Zhao, H.; Fu, L.; Cui, J.; Yang, Y. Global Trends and Research Progress of Photodynamic Therapy in Skin Cancer: A Bibliometric Analysis and Literature Review. *Clin. Cosmet. Investig. Dermatol.* **2023**, *16*, 479–498. [CrossRef]
31. Tanaka, Y.; Murayama, Y.; Matsumoto, T.; Kubo, H.; Harada, K.; Matsuo, H.; Kubota, T.; Okamoto, K.; Otsuji, E. Efficacy of 5-aminolevulinic acid-mediated photodynamic therapy in a mouse model of esophageal cancer. *Oncol. Lett.* **2020**, *20*, 82. [CrossRef] [PubMed]
32. Pedrosa, L.; Bedia, C.; Diao, D.; Mosteiro, A.; Ferrés, A.; Stanzani, E.; Martínez-Soler, F.; Tortosa, A.; Pineda, E.; Aldecoa, I.; et al. Preclinical Studies with Glioblastoma Brain Organoid Co-Cultures Show Efficient 5-ALA Photodynamic Therapy. *Cells* **2023**, *12*, 1125. [CrossRef]
33. Kamanlı, A.F.; Yıldız, M.Z.; Özyol, E.; Deveci Ozkan, A.; Sozen Kucukkara, E.; Guney Eskiler, G. Investigation of LED-based photodynamic therapy efficiency on breast cancer cells. *Lasers Med. Sci.* **2021**, *36*, 563–569. [CrossRef]
34. Sansaloni-Pastor, S.; Lange, N. Unleashing the potential of 5-Aminolevulinic acid: Unveiling a promising target for cancer diagnosis and treatment beyond photodynamic therapy. *J. Photochem. Photobiol. B* **2023**, *247*, 112771. [CrossRef] [PubMed]
35. Kumar, A.; Pecquenard, F.; Baydoun, M.; Quilbé, A.; Moralès, O.; Leroux, B.; Aoudjehane, L.; Conti, F.; Boleslawski, E.; Delhem, N. An Efficient 5-Aminolevulinic Acid Photodynamic Therapy Treatment for Human Hepatocellular Carcinoma. *Int. J. Mol. Sci.* **2023**, *24*, 426. [CrossRef] [PubMed]
36. Gondivkar, S.M.; Gadbail, A.R.; Choudhary, M.G.; Vedpathak, P.R.; Likhitkar, M.S. Photodynamic treatment outcomes of potentially-malignant lesions and malignancies of the head and neck region: A systematic review. *J. Investig. Clin. Dent.* **2018**, *9*, e12270. [CrossRef] [PubMed]
37. Kelloniemi, E.; Järvinen, R.; Hellström, P.; Rintala, E.; Aaltomaa, S.; Isotalo, T.; Innos, K.; Kaasinen, E. Repeated 5-aminolevulinic Acid Instillations During Follow-up in Non-muscle-invasive Bladder Cancer: A Randomized Study. *In Vivo* **2021**, *35*, 1561–1568. [CrossRef] [PubMed]
38. Kohoutova, D.; Haidry, R.; Banks, M.; Butt, M.A.; Dunn, J.; Thorpe, S.; Lovat, L. Long-term outcomes of the randomized controlled trial comparing 5-aminolaevulinic acid and Photofrin photodynamic therapy for Barrett's oesophagus related neoplasia. *Scand. J. Gastroenterol.* **2018**, *53*, 527–532. [CrossRef] [PubMed]
39. Su, Y.; Zhang, Y.; Tong, Y.; Zhang, L.; Li, P.; Zhang, H.; Zhang, X.; Tang, Y.; Qin, L.; Shen, Y.; et al. Effect and rational application of topical photodynamic therapy (PDT) with 5-aminolevulinic acid (5-ALA) for treatment of cervical intraepithelial neoplasia with vaginal intraepithelial neoplasia. *Photodiagnosis Photodyn. Ther.* **2022**, *37*, 102634. [CrossRef] [PubMed]
40. Bai-Habelski, J.C.; Ko, A.; Ortland, C.; Stocker, M.; Ebeling, A.; Reinhold, U. 5-ALA loaded self-adhesive patch-PDT is effective and safe in the treatment of actinic keratoses on hands and arms. *Exp. Dermatol.* **2022**, *31*, 1385–1391. [CrossRef] [PubMed]
41. Wang, L.; Chelakkot, V.S.; Newhook, N.; Tucker, S.; Hirasawa, K. Inflammatory cell death induced by 5-aminolevulinic acid-photodynamic therapy initiates anticancer immunity. *Front. Oncol.* **2023**, *13*, 1156763. [CrossRef]
42. Bhanja, D.; Wilding, H.; Baroz, A.; Trifoi, M.; Shenoy, G.; Slagle-Webb, B.; Hayes, D.; Soudagar, Y.; Connor, J.; Mansouri, A. Photodynamic Therapy for Glioblastoma: Illuminating the Path toward Clinical Applicability. *Cancers* **2023**, *15*, 3427. [CrossRef]
43. Regula, J.; Ravi, B.; Bedwell, J.; MacRobert, A.J.; Bown, S.G. Photodynamic therapy using 5-aminolaevulinic acid for experimental pancreatic cancer–prolonged animal survival. *Br. J. Cancer* **1994**, *70*, 248–254. [CrossRef]
44. Wang, X.; Jin, J.; Li, W.; Wang, Q.; Han, Y.; Liu, H. Differential in vitro sensitivity of oral precancerous and squamous cell carcinoma cell lines to 5-aminolevulinic acid-mediated photodynamic therapy. *Photodiagnosis Photodyn. Ther.* **2020**, *29*, 101554. [CrossRef] [PubMed]
45. Khaled, Y.S.; Wright, K.; Melcher, A.; Jayne, D. Anti-cancer effects of oncolytic viral therapy combined with photodynamic therapy in human pancreatic cancer cell lines. *Lancet* **2015**, *385* (Suppl. S1), S56. [CrossRef] [PubMed]
46. Lanuti, P.; Bertagnolo, V.; Pierdomenico, L.; Bascelli, A.; Santavenere, E.; Alinari, L.; Capitani, S.; Miscia, S.; Marchisio, M. Enhancement of TRAIL cytotoxicity by AG-490 in human ALL cells is characterized by downregulation of cIAP-1 and cIAP-2 through inhibition of Jak2/Stat3. *Cell Res.* **2009**, *19*, 1079–1089. [CrossRef] [PubMed]
47. Darzynkiewicz, Z.; Bedner, E.; Smolewski, P. Flow cytometry in analysis of cell cycle and apoptosis. *Semin. Hematol.* **2001**, *38*, 179–193. [CrossRef] [PubMed]
48. Telford, W.G.; Komoriya, A.; Packard, B.Z.; Bagwell, C.B. Multiparametric analysis of apoptosis by flow cytometry. *Methods Mol. Biol.* **2011**, *699*, 203–227. [CrossRef] [PubMed]
49. Bologna, G.; Lanuti, P.; D'Ambrosio, P.; Tonucci, L.; Pierdomenico, L.; D'Emilio, C.; Celli, N.; Marchisio, M.; d'Alessandro, N.; Santavenere, E.; et al. Water-soluble platinum phthalocyanines as potential antitumor agents. *Biometals* **2014**, *27*, 575–589. [CrossRef] [PubMed]
50. Moan, J.; Bech, O.; Gaullier, J.M.; Stokke, T.; Steen, H.B.; Ma, L.W.; Berg, K. Protoporphyrin IX accumulation in cells treated with 5-aminolevulinic acid: Dependence on cell density, cell size and cell cycle. *Int. J. Cancer* **1998**, *75*, 134–139. [CrossRef]
51. Jiang, W.; Liang, M.; Lei, Q.; Li, G.; Wu, S. The Current Status of Photodynamic Therapy in Cancer Treatment. *Cancers* **2023**, *15*, 585. [CrossRef] [PubMed]

52. Alvarez, N.; Sevilla, A. Current Advances in Photodynamic Therapy (PDT) and the Future Potential of PDT-Combinatorial Cancer Therapies. *Int. J. Mol. Sci.* **2024**, *25*, 1023. [CrossRef]
53. Mokoena, D.R.; George, B.P.; Abrahamse, H. Photodynamic Therapy Induced Cell Death Mechanisms in Breast Cancer. *Int. J. Mol. Sci.* **2021**, *22*, 10506. [CrossRef]
54. Chilakamarthi, U.; Giribabu, L. Photodynamic Therapy: Past, Present and Future. *Chem. Rec.* **2017**, *17*, 775–802. [CrossRef] [PubMed]
55. Koizumi, N.; Harada, Y.; Minamikawa, T.; Tanaka, H.; Otsuji, E.; Takamatsu, T. Recent advances in photodynamic diagnosis of gastric cancer using 5-aminolevulinic acid. *World J. Gastroenterol.* **2016**, *22*, 1289–1296. [CrossRef] [PubMed]
56. Klein, A.; Babilas, P.; Karrer, S.; Landthaler, M.; Szeimies, R.M. Photodynamic therapy in dermatology—An update 2008. *J. Dtsch. Dermatol. Ges.* **2008**, *6*, 839–845. [CrossRef] [PubMed]
57. Maharjan, P.S.; Bhattarai, H.K. Singlet Oxygen, Photodynamic Therapy, and Mechanisms of Cancer Cell Death. *J. Oncol.* **2022**, *2022*, 7211485. [CrossRef] [PubMed]
58. Girotti, A.W.; Fahey, J.M.; Korbelik, M. Photodynamic Therapy as an Oxidative Anti-Tumor Modality: Negative Effects of Nitric Oxide on Treatment Efficacy. *Pharmaceutics* **2021**, *13*, 593. [CrossRef] [PubMed]
59. Chen, X.; Zhao, P.; Chen, F.; Li, L.; Luo, R. Effect and mechanism of 5-aminolevulinic acid-mediated photodynamic therapy in esophageal cancer. *Lasers Med. Sci.* **2011**, *26*, 69–78. [CrossRef] [PubMed]
60. Rosin, F.C.P.; Teixeira, M.G.; Pelissari, C.; Corrêa, L. Photodynamic Therapy Mediated by 5-aminolevulinic Acid Promotes the Upregulation and Modifies the Intracellular Expression of Surveillance Proteins in Oral Squamous Cell Carcinoma. *Photochem. Photobiol.* **2019**, *95*, 635–643. [CrossRef] [PubMed]
61. Zhu, M.Q.; Shi, S.R.; Wan, G.Y.; Wang, Y.S.; Wang, Y.; Zhang, L.Y.; Zhao, Y.H. Study on the in vivo and in vitro effects of photodynamic therapy mediated by 5-aminolevulinic acid on oral squamous cell carcinoma. *Zhonghua Kou Qiang Yi Xue Za Zhi* **2019**, *54*, 176–182. [CrossRef] [PubMed]
62. Wang, X.; Li, N.; Meng, J.; Wen, N. The use of topical ALA-photodynamic therapy combined with induction chemotherapy for locally advanced oral squamous cell carcinoma. *Am. J. Otolaryngol.* **2021**, *42*, 103112. [CrossRef]
63. Abd-Elgaliel, W.R.; Cruz-Monserrate, Z.; Wang, H.; Logsdon, C.D.; Tung, C.H. Pancreatic cancer-associated Cathepsin E as a drug activator. *J. Control. Release* **2013**, *167*, 221–227. [CrossRef]

biomedicines

MDPI

Review

Potential of Resveratrol to Combine with Hydrogel for Photodynamic Therapy against Bacteria and Cancer—A Review

Siu Kan Law *,†, Cris Wai Ching Liu †, Christy Wing Sum Tong and Dawn Ching Tung Au *

Department of Food and Health Sciences, The Technological and Higher Education Institute of Hong Kong, Tsing Yi, New Territories, Hong Kong, China; criswcliu@thei.edu.hk (C.W.C.L.); christytong@thei.edu.hk (C.W.S.T.)

* Correspondence: siukanlaw@thei.edu.hk or siukanlaw@hotmail.com (S.K.L.); dawnau@thei.edu.hk (D.C.T.A.)

† These authors contributed equally to this work.

Abstract: Bacterial infections and cancers are important issues in public health around the world. Currently, Western medicine is the most suitable approach when dealing with these issues. "Antibiotics" and "Corticosteroids" are the Western medicines used for bacterial infection. "Chemotherapy drugs", "surgery", and "radiotherapy" are common techniques used to treat cancer. These are conventional treatments with many side effects. PDT is a non-invasive and effective therapy for bacterial infection and cancer diseases. Methods: Nine electronic databases, namely WanFang Data, PubMed, Science Direct, Scopus, Web of Science, Springer Link, SciFinder, and China National Knowledge Infrastructure (CNKI), were searched to conduct this literature review, without any regard to language constraints. Studies focusing on the photodynamic actions of hydrogel and Resveratrol were included that evaluated the effect of PDT against bacteria and cancer. All eligible studies were analyzed and summarized in this review. Results: Resveratrol has antibacterial and anticancer effects. It can also act as PS in PDT or adjuvant but has some limitations. This is much better when combined with a hydrogel to enhance the effectiveness of PDT in the fight against bacteria and cancer. Conclusions: Resveratrol combined with hydrogel is possible for PDT treatment in bacteria and cancer. They are compatible and reinforce each other to increase the effectiveness of PDT. However, much more work is required, such as cytotoxicity safety assessments of the human body and further enhancing the effectiveness of PDT in different environments for future investigations.

Keywords: resveratrol; hydrogel; photodynamic therapy; bacteria; cancer

1. Introduction

Bacterial infection and cancer disease are issues involved in public health around the world. Currently, Western medicine is the common approach to managing these conditions.

Antibiotics and corticosteroids are medicines that are used to fight bacterial infections. They either kill or stop bacterial growth by affecting self-reproduction and preventing mutation, gene recombination, and transfer. This allows the body's natural defenses to eliminate the pathogens [1]. The negative impacts of antibiotics are increasing morbidity and mortality, leading to antimicrobial resistance (AMR), which poses a major threat to human health in the world [2].

Chemotherapy drugs, surgery, and radiotherapy are the general techniques used to fight against cancer, but these have side effects on the human body [3]. Therefore, photodynamic therapy (PDT) has become more popular in medical usage.

PDT is a non-invasive therapy, which is different from traditional surgery. This is a selective and repeatable treatment, unlike radiation therapy, which is harmful to normal cells or tissues. Compared with antibiotics and corticosteroid chemotherapy, PDT is more effective and has fewer side effects [4].

Resveratrol is a natural product that has been extracted from Chinese herbal plants [5], and it has been applied for the treatment of different diseases, such as bacterial infections and cancer. It also can act as a photosensitizer (PS) in PDT therapy or adjuvant. If it is a PS, a photo-induced chemical reaction of *trans*-Resveratrol occurs to generate singlet oxygen ($^1O_2$) during the PDT process [6].

PDT has been used for a long time in the treatment of bacterial infections and cancer. Antibacterial photodynamic therapy (aPDT) has the potential to kill multidrug-resistant pathogenic bacteria, such as Gram-positive bacteria (*Staphylococcus aureus* and *Enteroccoccus fecalis*) and Gram-negative bacteria (*Escherichia coli, Proteus mirabilis*, and *Pseudomonas aeruginosa*). It has a low tendency to induce drug resistance, and bacteria rapidly develop against traditional antibiotic therapy [7]. PDT enables the treatment of multifocal disease with the least amount of tissue damage in terms of cancer [8]. The mechanisms of PDT, either in bacterial infection or cancer, are quite similar and are involved in Types I and II to generate the singlet oxygen ($^1O_2$) and reactive oxygen species (ROS) in the electron-transfer reaction.

Resveratrol can act as a natural PS, but it has some limitations for PDT, for example, its short wavelength and poor bioavailability. It is usually combined with hydrogel to enhance the function of PDT. Resveratrol and hydrogel are complementary during the PDT process in terms of maximizing effectiveness. Recently, Gan and Liu et al. have published a similar review of hydrogel-based phototherapy and drug-delivery systems [9,10], but these do not describe the natural product Resveratrol.

Hence, this review article is divided into five parts, namely (1) the definition of hydrogel, (2) the theory and mechanisms of PDT, (3) natural PS, especially in the case of Resveratrol, (4) the photodynamic action of hydrogel, and (5) the possibility of combining hydrogel with Resveratrol for PDT against bacteria and cancers.

1.1. Hydrogels

Hydrogels are three-dimensional (3D) network structures with crosslinked polymer chains that form ionic or covalent bonds for linking one polymer chain to another [11]. Glutaraldehyde [12,13] is the most common compound for crosslinking a hydrogel, and other crosslinking compounds include Formaldehyde [14], Epoxy [15], etc. (Figure 1). The crosslinking changes a liquid polymer into a solid or gel by restricting its ability to move. Since the crosslinking structure of a hydrogel can absorb relatively large amounts of water, its properties are soft and resemble living tissues in humans [16].

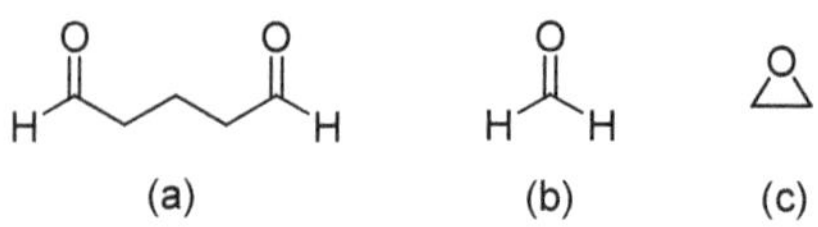

Figure 1. Crosslinking compounds of hydrogel: (**a**) Glutaraldehyde, (**b**) Formaldehyde, and (**c**) Epoxy.

Hydrogels are porous, with spaces available between adjacent crosslinking structures in the polymer network [17]. There are some development theories for porous hydrogels, including the kinetics of swelling, equilibrium swelling, the structure–stiffness relationship, and solute diffusion in dense hydrogels.

The swelling of a hydrogel is the kinetic process of coupling mass transport and mechanical deformation that includes linear and non-linear poroelasticity. These theories are consistent within the linear regime under small perturbations from an isotropically swollen hydrogel state, such as a change in the volume of a hydrogel as it absorbs a compatible solvent or water content [18]. Hydrogel also causes equilibrium swelling depending on the pH of an environment. It acts as a polyampholyte polyelectrolyte when immersed in an ionic solution, producing hydrogen and hydroxide ions in a reaction [19].

This is the soft or weak material with a structure–stiffness relationship. Water content and mechanical properties of the hydrogel are adjusted over a wide range through the dissociation and reformation of hydrogen bonds by solvent exchange and the heating process [20]. Diffusion of solutes within the hydrogel occurs through nano-to-microscopic open spaces or dynamic free volumes between the aqueous solution and the liquid-filled polymer fibers, which includes sub-nanometer-scale cavities between molecules [21].

1.2. Examples of Antibacterial and Anticancer Applications

Hydrogels are naturally derived from alginate and chitosan or synthetic modification from polyacrylamide. Fasiku et al. reported a chitosan-based hydrogel for dual delivery with hydrogen peroxide of antimicrobial peptide against bacterial methicillin-resistant *Staphylococcus aureus* biofilm-infected wounds that were prepared through the Michael addition technique (Figure 2) [22]. In 2022, Abbasalizadeh et al. developed a curcumin-chrysin-alginate-chitosan hydrogel that was prepared through the ionic gelation mechanism utilizing $CaCl_2$ to treat breast (T47D) and lung cancers (A549) (Figure 3) [23]. Lu et al. identified the polyacrylamides hydrogel causing *Staphylococcus aureus* and *Escherichia coli* differentiation upon visible light irradiation (Figure 4) [24]. Andrade et al. indicated that stimuli-responsive hydrogel was able to change its physical state from liquid to gel according to external factors such as temperature, pH, light, ionic strength, and magnetic field for cancer treatment (Figure 5) [25].

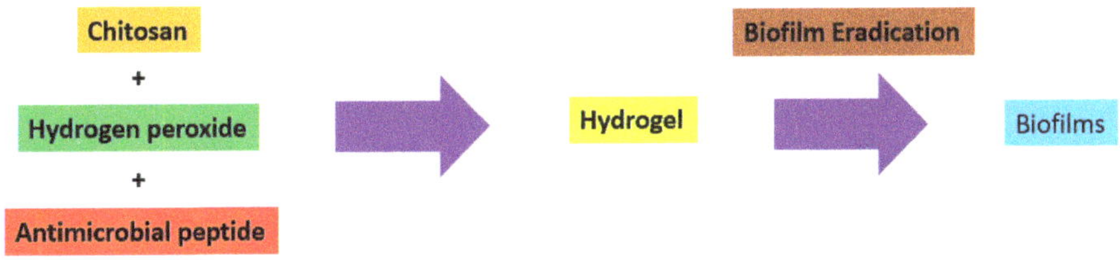

Figure 2. Synthetic diagram for the synthesis of a chitosan-based hydrogel with hydrogen peroxide of antimicrobial peptide against *Staphylococcus aureus*.

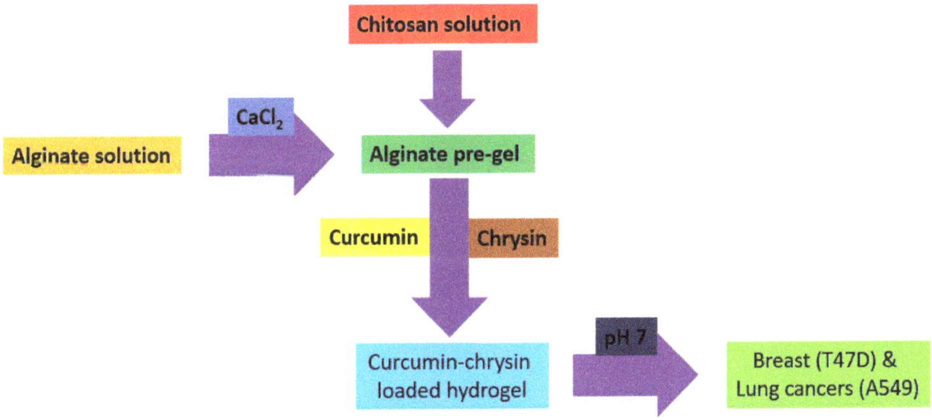

Figure 3. Synthetic diagram for the synthesis of a curcumin-chrysin-alginate-chitosan hydrogel against breast (T47D) and lung cancers (A549).

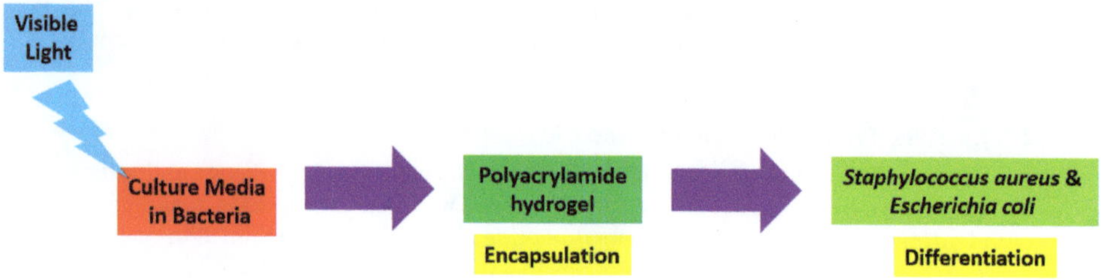

Figure 4. Synthetic diagram for the synthesis of a polyacrylamide hydrogel against *Staphylococcus aureus* and *Escherichia coli*.

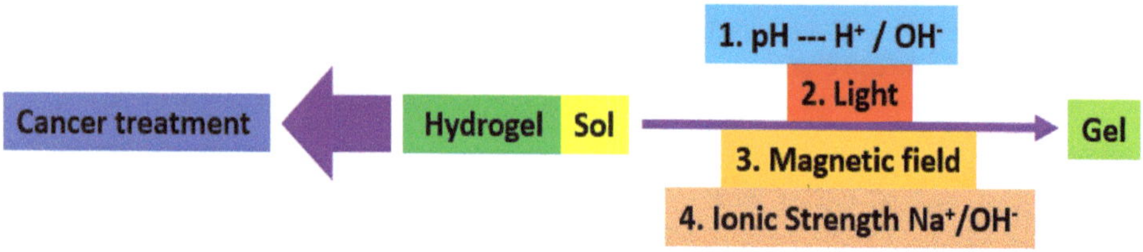

Figure 5. Synthetic diagram for the synthesis of a stimuli−responsive hydrogel for cancer treatment.

1.3. Drug-Delivery System

There are some development theories of hydrogels to the drug-delivery approach. First, hydrogel establishes scaffolds to address easily inadequate local drug availability and delivery sites. These have a highly hydrated mesh network formed from natural, synthetic, or semi-synthetic polymers with physically or covalently crosslinked (Table 1) [26]. They provide high biocompatibility [27], drug protection [28], spatiotemporal control of the drug release [29], and physicochemical tailorability [30].

Table 1. Classification of hydrogel for the formation of crosslinking (modified from [10]).

Crosslinking Methods	Physical crosslinking	Hydrogen bonding
		Electrostatic interaction
		Van der Waals forces
		Host-guest interactions
		Crystallization
	Chemical crosslinking	Free radical reaction
		Carbodiimide chemistry
		Click chemistry
		Enzyme-mediated reaction
		Condensation polymerization

Biocompatibility is the ability of a biomaterial (e.g., hydrogel) to perform with an appropriate host response in the specific application [27]. Hydrogels also serve as a platform on which various physiochemical interactions with the encapsulated drugs occur to control drug release [28]. This provides a wide range of drug–carrier interface modifications to maximize drug-delivery needs through physical (electric field, magnetic field, light), mechanical (ultrasound, mechanical strain), or chemical (pH, redox gradient, enzymes), and without side effects [29].

Moreover, the physicochemical tailorability sparks through the inherent bioactivity of gelatin in tissue engineering. Spatial and temporal control of hydrogel crosslinking and its rheological behavior enables deposition via a variety of manufacturing-related techniques. Hydrogels can tailor to the remodeling rate of the target tissue to produce viable well-defined 3D structures or long-term cellular performance in non-biofabricated structures [30]. Therefore, hydrogel is a suitable material as a drug-delivery system applied in PDT.

1.4. Reasons for Hydrogel Use in PDT

In general, hydrogel has several competitive advantages, including good biocompatibility and low cytotoxicity, enhanced antitumor effect, reduced toxic and side effects, as well as maintaining a high concentration and archive the selective of a drug during the PDT process. These characteristics of hydrogel overcome the limitations of PDT and enhance its effectiveness, which has been discussed in drug-delivery systems. Previous studies have investigated a novel hydrogel shell on cancer cells that were prepared through in situ photopolymerization of polyethyleneglycol diacrylate (PEGDA) using methylene blue (MB)-sensitized mesoporous titanium oxide (TiO_2) nanocrystal for enhancing the effectiveness of PDT. TiO_2 was the PS and activated the formation of hydrogel, which protected the MB and acted as a significant photosensitive additive to improve the treatment of PDT. MB would be eliminated and inactivated after undergoing the PDT process [31]. Thus, there is a relationship between the principle of PDT and hydrogel.

2. Principle of PDT

The basic principle of PDT is a dynamic interaction between the photosensitizer (PS) and light with a specific wavelength that generates the reactive oxygen species (ROS), such as singlet oxygen ($^1O_2$) and superoxide anion (O_2^-) that causes irreversible oxidative damage to the cell membrane and DNA, leading to the selective destruction of target tissue, bacteria, and cancer [32].

Conventional PDT with the advantages of direct targeting, low invasiveness, remarkable curative, and fewer side effects [33]. However, it has limited or weak light-penetration depth, a short release distance, and a lifetime of ROS to fight against bacteria [34]. This also exists alongside poor selectivity and oxygen dependence for the therapeutic effectiveness of solid tumors [35]. Table 2 summarizes the representative cases of recent progress in PDT against bacteria and cancer over five years.

Table 2. Recent progress in PDT against bacteria and cancer.

	Bacteria		
	Study	**Consequences**	**References**
1	Type-I photodynamic antimicrobial therapy: Principles, progress, and future perspectives	The fundamental principles of Type-I PDT were discussed, including its physicochemical properties and the generation of ROS, as well as explored several specific antimicrobial mechanisms utilized by Type-I PDT and summarized the recent applications of Type-I PDT in antimicrobial treatment.	[36]
2	Photodynamic therapy for the treatment of *Pseudomonas aeruginosa* infections: A scoping review	PDT was an effective adjunct to antimicrobial therapy against *Pseudomonas aeruginosa*, according to the usage of PS and the infection location, but the evidence was supported significantly by in vitro than the in vivo studies.	[37]
3	Antibacterial Photodynamic Therapy in the Near-Infrared Region with a Targeting Antimicrobial Peptide Connected to a π-Extended Porphyrin	Antimicrobial PDT upon irradiation at 720 nm for the conjugation consisted of an antimicrobial peptide linked to a π-extended porphyrin photosensitizer, which was at micromole concentration with a strong effect on both Gram-positive and Gram-negative bacteria.	[38]

Table 2. *Cont.*

Cancer			
	Study	Consequences	References
4	Recent Progress and Trends in X-ray-Induced Photodynamic Therapy with Low Radiation Doses	The concept of X-PDT and its relationships with radiodynamic therapy and radiotherapy, as well as the mechanism of X-ray absorption and conversion by scintillating materials, ROS evaluation for low dosage X-PDT, radiation side effects, and clinical concerns on X-ray radiation, were discussed.	[39]
5	Progress in Clinical Trials of Photodynamic Therapy for Solid Tumors and the Role of Nanomedicine	The latest clinical studies and preclinical in vivo studies on the use of PDT and its progress on nano-therapeutics as delivery tools for PS, which improved their cancer cellular uptake and their toxic properties	[40]
6	Trial watch: an update of clinical advances in photodynamic therapy and its immunoadjuvant properties for cancer treatment	Trial Watch provided recent clinical information on the immunomodulatory properties of PDT and ongoing clinical trials using PDT to treat cancer patients.	[41]

2.1. Mechanisms

In general, PDT causes the activation of PS through visible or ultraviolet light, which generates the $^1O_2$ and free radicals. The electrons of the PS are promoted from the ground to an excited state by the light activation. These excited singlet-state electrons from the PS are unstable and release the excess energy backing to the ground state by emitting fluorescence. Its excited singlet-state electrons can also change and shift to the triplet state through the intersystem crossing system [42].

There are two types of pathways for the triplet state of PS that react with the substrate [43]. Type-I and II pathways involve the electron transfer from triplet state PS. The difference between these two pathways is the generation of final products. Type-I pathway consists of free radicals, and these interact with oxygen at a molecular level to form ROS, such as superoxide ($O_2^{\bullet-}$), hydroxyl radicals (•OH), and hydrogen peroxide (H_2O_2). Type II pathway contains $^1O_2$ only generated by the electron transfer between the excited PS and the ground-state molecular oxygen [44]. Type-I pathway operates in an oxygen (O_2)-independent manner, and the type II pathway relies on the presence of molecular oxygen (Figure 6).

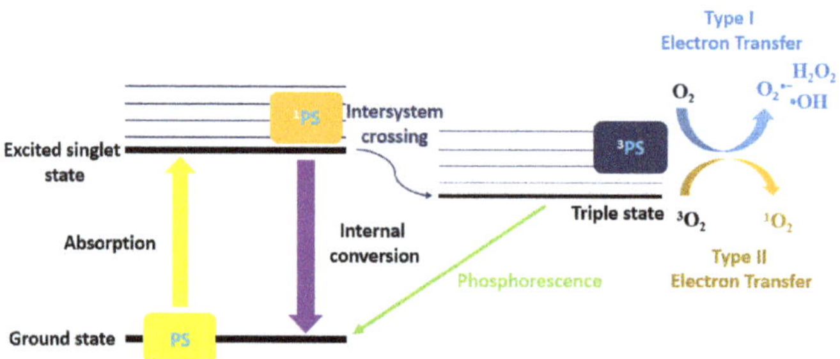

Figure 6. Schematic diagram of the PDT Types I and II mechanisms.

These Type-I and II pathways of antimicrobial photodynamic therapy (aPDT) are employed for killing Gram-positive bacteria (*Staphylococcus aureus*, *Enteroccoccus fecalis*) and Gram-negative bacteria (*Escherichia coli*, *Proteus mirabilis*, and *Pseudomonas aeruginosa*) bacteria [42], which causes irreversible oxidative damage to the cell membrane and DNA, leading to bacterial cell death [45]. Both Type-I and II pathways are also applied in the treatment of cancers that can trigger different cell death mechanisms and are directly cytotoxic to the cancer cells resulting in apoptosis or necrosis [46]. The percentage or ratio of the final products for Type-I and II pathways depends on the nature or type of PS.

2.2. Natural Photosensitizer (PS)

Typically, PS molecules absorb light with a suitable wavelength that initiates the activation processes for the selective destruction of cancer cells [47]. Many natural PS from traditional Chinese medicinal plants are applied in PDT, including Pheophorbide a (Pa) [48], Hypocrellin B (HB) [49], and Curcumin (Cur) (Figure 7) [50], since these are phytochemical compounds with less- or non-toxic side effects [51].

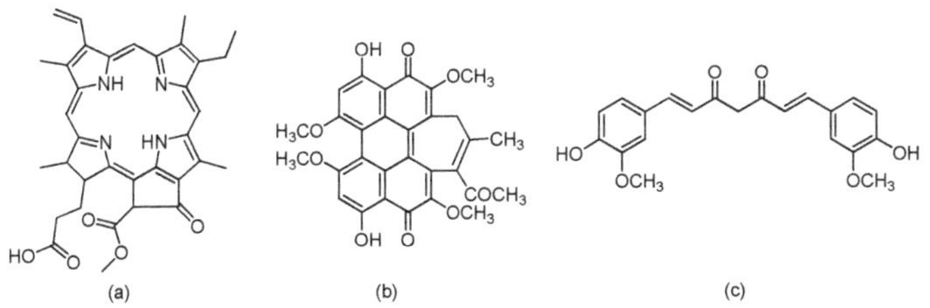

Figure 7. Chemical structures of (**a**) Pheophorbide a (Pa), (**b**) Hypocrellin B (HB), and (**c**) Curcumin (Cur).

They usually act as antibacterial and anticancer agents of natural products origin [52]. PS can selectively target microbes and leave out normal tissue, making it more efficient against the bacterial infection site when as an antibacterial agent. If the PS is an anticancer agent, it reduces toxicity to healthy tissues and has a lower incidence of side effects [53]. Some representative examples of natural PS for PDT against bacteria and cancer are shown in Table 3. Resveratrol as a PS will be discussed in the next section.

Table 3. Natural PS for PDT against bacteria and cancer.

		Bacteria			
	Study	**Natural Photosensitizer and Dosage**	**Usage of Light and Energy (J)**	**Consequences**	**Reference**
1	Bactericidal Effect of Photodynamic Therapy Using Na-Pheophorbide a: Evaluation of Adequate Light Source	280 μM of Na-Pheophorbide a on *Staphylococcus aureus*.	Irradiated with red light at 670 nm in 300 W/cm^2, and fluence 27 J/cm^2 for 30 min.	PDT with Na-Pheophorbide a possessed a better anti-bactericidal function that was useful for the treatment of septic arthritis and soft tissue infection.	[54]
2	Hypocrellin B-Mediated Photodynamic Inactivation of Gram-Positive Antibiotic-Resistant Bacteria: An in vitro Study	100 μM of Hypocrellin B on *Staphylococcus aureus*, *Enterococcus faecalis*, *Streptococcus pneumonia*, *Escherichia coli*, and *Klebsiella pneumoniae*.	Irradiated with red light at 660 nm in 0.5 W/cm^2, and fluence 72 J/cm^2 for 30 min.	PDT with Hypocrellin B was effective in inactivating the Gram-positive bacteria, including *Staphylococcus aureus*, *Enterococcus faecalis*, and *Streptococcus pneumonia* bacteria.	[55]
3	Antimicrobial photodynamic therapy with curcumin on methicillin-resistant *Staphylococcus aureus* biofilm	80 μM of Curcumin on *Staphylococcus aureus*	Irradiated with blue light at 450 nm in 110 W/cm^2, and fluence 50 J/cm^2 for 455 s.	PDT with curcumin reduced the biofilm viability of *Staphylococcus aureus*, attesting to the efficiency of the therapy, which was internalized by bacterial cells even in biofilm aggregates,	[56]

Table 3. *Cont.*

Cancer					
	Study	**Natural Photosensitizer and Dosage**	**Usage of Light and Energy (J)**	**Consequences**	**Reference**
1	Pheophorbide a-Mediated Photodynamic Therapy Triggers HLA Class I-Restricted Antigen Presentation in Human Hepatocellular Carcinoma	0.35 μM of Pheophorbide a on Human Hepatocellular Carcinoma, HepG2.	Irradiated with red light at 670 nm in 70 W/cm^2, and fluence 84 J/cm^2 for 20 min.	PDT with Pheophorbide a triggered phagocytic capture by human macrophages, causing apoptosis and cancer immunity in the tumor host.	[48]
2	Effect of photodynamic therapy with Hypocrellin B on apoptosis, adhesion, and migration of cancer cells	2.5 μM of Hypocrellin B on Human ovarian cancer, HO-8910.	Irradiated with red light at 660 nm in 0.5 W/cm^2, and fluence 72 J/cm^2 for 5 h.	PDT with Hypocrellin B induced apoptosis and inhibited adhesion and migration of cancer cells in vitro.	[57]
3	Assessing the Effects of Curcumin and 450 nm Photodynamic Therapy on Oxidative Metabolism and Cell Cycle in Head and Neck Squamous Cell Carcinoma: An in vitro Study	0.1 to 10 μM of Curcumin on Human Head and Neck Squamous Cell Carcinoma, HNSCC.	Irradiated with blue light at 450 nm in 0.25 W/cm^2, and fluence 60 J/cm^2 for 1 h.	PDT with Curcumin increased oxidative damage, reduced cellular growth, and a cell cycle block in the G1 phase for Human Head and Neck Squamous Cell Carcinoma.	[58]

2.2.1. Resveratrol (3,5,4′-Trihydroxystilbene) (RSV)

This is a plant compound with a natural polyphenol stilbene structure that is isolated from the root of *Veratrum grandiflorum* [59] or *darakchasava*, or *manakka* for medicinal purposes [60]. Its molecular weight of 228.25 g/mol consists of two phenolic rings formed by a double styrene bond. Thus, there are two isomeric *cis*- and *trans*-forms (Figure 8).

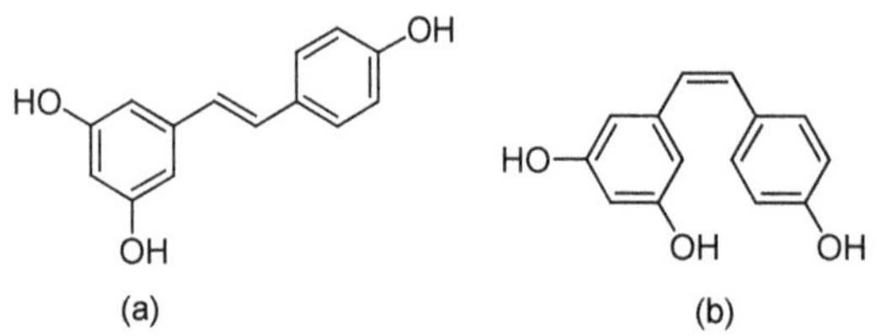

Figure 8. Chemical structures of (**a**) *trans*-Resveratrol, and (**b**) *cis*-Resveratrol.

The *cis*- and *trans*- forms of Resveratrol can co-exist in a variety of fruits, including grapes (Vitaceae) [61] and blueberries (*Vaccinium* spp.), blackberries (*Morus* spp.), and peanuts (*Arachis hypogaea*) [62,63], and red wine. *Trans*-Resveratrol is the predominant form and is more stable than *cis*-Resveratrol in nature [64,65]. It is a bioactive compound with a wide range of pharmacology properties, e.g., immunomodulatory [66], glucose and lipid regulatory [67], neuroprotective [68], cardiovascular protective effects [69], antioxidant [70], anti-inflammatory [71], antibacterial [72–77], and anticancer (Table 4) [78–87].

Table 4 shows the investigation of an antibacterial and anticancer effect on Resveratrol. The antibacterial mechanisms of Resveratrol consist of DNA damage [88], cell division impairment [89], oxidative membrane damage [90], and metabolic enzyme inhibition [91].

Table 4. Antibacterial and anticancer effects of Resveratrol.

	Bacteria				
	Study	Models	Dosage and Time	Consequences	Reference
1	Resveratrol antibacterial activity against Escherichia coli is mediated by Z-ring formation inhibition via suppression of FtsZ expression	*lacZ*$^+$ gene of an SOS-inducible sulA promoter on *Escherichia coli.*	456 μM of Resveratrol for 6 h.	Resveratrol increased DNA fragmentation and the expression level of SOS response-related genes in a dose-dependent manner, which inhibits bacterial cell growth by suppressing FtsZ expression and Z-ring formation.	[72]
2	Chemically Tuning Resveratrol for the Effective Killing of Gram-Positive Pathogens	*Bacillus cereus, Clostridium strains, Clostridioides difficile, Enterococcus faecalis, Streptococcus aureus*	40 to 160 μM of Resveratrol and its derivatives for 8 h.	Resveratrol and its derivatives with bactericidal activity against Gram-positive bacteria in the same low micromolar range of traditional antibiotics for disturbing the membrane permeability.	[73]
3	The Antibacterial and Antibiofilm Activities of Resveratrol on Gram-positive and Gram-negative Bacteria	*Staphylococcus aureus, Bacillus subtilis, Escherichia coli*, and *Pseudomonas aeruginosa.*	0.5 to 16 μM of Resveratrol for 24 h.	Resveratrol with antibacterial effect on Gram-positive and Gram-negative bacteria reduced the toxin production and inhibited biofilm formation.	[74]
4	Resveratrol enhances the antimicrobial effect of polymyxin B on Klebsiella pneumoniae and Escherichia coli isolates with polymyxin B resistance	6 strains of *Klebsiella pneumoniae* and 24 strains of *Escherichia coli.*	4 μM and 2 μM of polymyxin B for 6 strains of *Klebsiella pneumoniae* and 24 strains of *Escherichia coli* with an additional 32 to 128 μM Resveratrol for 24 h.	Resveratrol has increased the sensitivity of bacterial strains to polymyxin B, which is suitable for the treatment of bacterial infections.	[75]
5	Resveratrol Reverts Tolerance and Restores Susceptibility to Chlorhexidine and Benzalkonium in Gram-Negative Bacteria, Gram-Positive Bacteria, and Yeasts	*Enterococcus faecium, Staphylococcus aureus, Klebsiella pneumoniae, Acinetobacter baumannii, Pseudomonas aeruginosa* and *Enterobacter* spp.	32 to 256 μM of Resveratrol with 0.06 to 1024 μM of chlorhexidine and benzalkonium for 18 to 24 h.	Resveratrol reduced the dosage of chlorhexidine and benzalkonium, which was useful for biocides in several nosocomial pathogens.	[76]
6	Resveratrol-Induced Xenophagy Promotes Intracellular Bacteria Clearance in Intestinal Epithelial Cells and Macrophages	Life-threatening diseases, *Salmonella Typhimurium* and Crohn's disease-associated Adherent-Invasive *Escherichia coli.*	10 μM of Resveratrol for 20 h	Resveratrol stimulated xenophagy and enhanced the clearance of two invasive bacteria, including *Salmonella Typhimurium* and *Escherichia coli.*	[77]

Table 4. *Cont.*

	Cancer				
	Study	**Models**	**Dosage and Time**	**Consequences**	**Reference**
1	Resveratrol triggers autophagy-related apoptosis to inhibit the progression of colorectal cancer via inhibition of FOXQ1	Female BALB/c nude mice and C57BL/6 mice aged 6–8 weeks, or SW480-derived cancer cells.	Concentration between 50 and 80 μM for 48 h.	Resveratrol enhanced autophagy-related cell apoptosis in colorectal cancer cells through the SIRT1/FOXQ1/ATG16L pathway	[78]
2	Synergistic anticancer activity of Resveratrol in combination with docetaxel in prostate carcinoma cells	Prostate carcinoma LNCaP cells.	20 μM and 2 μM docetaxel for 24, 48, and 72 h.	Resveratrol improved the docetaxel therapy, targeting apoptosis and necroptosis simultaneously in the treatment of cancer	[79]
3	Metformin enhances anticancer properties of Resveratrol in MCF-7 breast cancer cells via induction of apoptosis, autophagy, and alteration in cell cycle distribution	Breast cancer cells, MCF-7.	2.5, 10, 25 mM of metformin, 70, 144, 287, 383, 575 μM of Resveratrol, and 20, 40 μM of cisplatin for 48 h.	Synergistic anticancer effects of metformin in a triple combination with cisplatin and Resveratrol were attributed to induction of autophagy-mediated cell death and apoptosis along cell cycle arrest	[80]
4	Resveratrol enhanced the anticancer effects of cisplatin on non-small cell lung cancer cell lines by inducing mitochondrial dysfunction and cell apoptosis	Human Lung cancer cells, H838 and H520.	40 μM and 55 μM of Resveratrol with 5 μM in H838 and H520 for 24 h.	Resveratrol exhibited its anticancer effects on non-small cell lung cancer H838 and H520 cell lines and enhanced the antitumor effects of cisplatin by regulating the mitochondrial apoptotic pathway.	[81]
5	Resveratrol Modulates the Redox-status and Cytotoxicity of Anticancer Drugs by Sensitizing Leukemic Lymphocytes and Protecting Normal Lymphocytes	Human Leukemic cells, U937 and HL-60.	12.5 μM of Resveratrol with 0.05 μM or 0.01 μM of barasertib, and 5 μM of everolimus.	In leukemic lymphocytes treated with barasertib and everolimus in the presence of Resveratrol, synergistic cytotoxicity was accompanied by strong induction of apoptosis, increased levels of hydroperoxides, and insignificant changes in protein-carbonyl products.	[82]
6	Efficacy of Resveratrol against breast cancer and hepatocellular carcinoma cell lines	Breast cancer cells MCF-7, and Hepatoblastoma cells, HepG2.	100 μM of Resveratrol h in MCF-7 and HepG2 for 24 h.	Resveratrol had a significant cytotoxic effect on MCF-7 and HepG2, which elevated caspase-3, caspase-8, caspase-9, Bax, p53, and p21 and reduced Bcl-2 and Bcl-xL mRNA levels in treating breast and liver cancers.	[83]

Table 4. *Cont.*

	Cancer				
	Study	Models	Dosage and Time	Consequences	Reference
7	Apoptotic effects of Resveratrol, a grape polyphenol, on imatinib-sensitive and resistant K562 chronic myeloid leukemia cells	Chronic myeloid leukemia cells K562 and K562/IMA-3 cells.	85 µM and 122 µM of Resveratrol in K562 and K562/IMA-3 cells for 72 h.	Resveratrol with antiproliferative and apoptotic effects on K562 cells, as well as increased more 100-folds in K562/IMA-3 cells.	[84]
8	Synergistic anticancer effects of curcumin and Resveratrol in Hepa1-6 hepatocellular carcinoma cells	Hepatocellular carcinoma cells, Hepa1-6.	40 µM and 10 µM of Resveratrol and curcumin for 48 h.	Resveratrol and curcumin induced the apoptosis of Hepa1-6 cells through the caspase-3, -8, and -9 activation, upregulated intracellular ROS levels, downregulated X-Linked Inhibitor of Apoptosis Protein (XIAP) and surviving expression.	[85]
9	In vivo Anticancer Effects of Resveratrol Mediated by NK Cell Activation	Natural killer cells, NK.	20 µM of Resveratrol and 5 µM of IL-2 for 36 h.	Resveratrol activated NK cells and synergistically increased IFN-γ secretion and NK cell cytotoxicity with interleukin-2 (IL-2), which inhibited tumor growth and metastasis in mice.	[86]
10	Celastrol and Resveratrol Modulate SIRT Genes Expression and Exert Anticancer Activity in Colon Cancer Cells and Cancer Stem-like Cells	Metastatic colon cancer LoVo cells and cancer stem-like cells LoVo/DX.	1.0 to 5.0 µM of Resveratrol and 1.0 to 5.0 µM of celastrol for 24 h.	Resveratrol and celastrol exerted an antitumor activity against metastatic LoVo cells and cancer stem-like LoVo/DX cells, while Resveratrol with a greater effect on colon cancer cells and less aggressive forms of the disease than celastrol.	[87]

Resveratrol increases the expression of genes from a SOS–stress regulon, which causes DNA fragmentation, cell cycle arrest, and DNA impairment in bacteria, such as *Escherichia coli* [72,88]. It can also affect the prokaryotic cell division protein (FtsZ) and GTPase activity. This protein is highly conserved throughout eubacteria, which may cause cell division impairment in bacteria [92].

The effect of Resveratrol on *E. coli* is not through diffusible ROS. It is mediated site-specifically as the primary event for oxidative damage of the cell membrane [90]. It also inhibits ATPase and the synthesis of ATP bound in a hydrophobic pocket between the C-terminal tip of the gamma subunit and the hydrophobic interior of the surrounding torus, a region critical for gamma-subunit rotation [91].

Resveratrol has several anticancer mechanisms, including initiating the apoptosis of mitochondria in the cytoplasm, enhancing oxidative stress, and interfering with energy metabolism in cancer cells. The generation of ROS causes cancer cell damage and death [93].

Mitochondria is an intrinsic apoptotic pathway causing stimulation through the activation of the proapoptotic Bcl-2 family (Bak, Bax). By suppressing the anti-apoptotic proteins Bcl-xL, Bcl-2, and Mcl-1, it diminished the potential of mitochondrial outer membrane permeability (MOMP). The cytochrome c then binds to this cytosolic apoptotic protease activating factor-1 (Apaf-1) [94], which recruits the initiator procaspase-9 activating downstream executor caspase-3, -6, and -7 for cleavage of cytoplasm leading to the cell apoptosis [95].

The intervention is an extrinsic apoptotic pathway involving tumor necrosis factor (TNF) to interact with cell surface receptors such as TRAILR1 (DR4)-TRAIL, TRAILR2 (DR5)-TRAIL, TNFR1-TNFα and FAS (CD95, APO-1)-FasL [96]. It causes apoptosis by the activation of caspase-3, -6, and -7, which leads to the disassembly of cell survival and death [97].

Meanwhile, Resveratrol inhibits the neurogenic locus notch homolog protein 1 to increase the generation of ROS. These signal transductions upregulate the phosphatase tensin homolog and down-regulate AKT Serine/Threonine Kinase 1. As a result, caspase-3 cleavage by decreasing phospho-Akt induced cell death [98]. Growing evidence has indicated that *trans*-Resveratrol has anticancer properties. It is a PS applied to the PDT to enhance its functions against bacteria and cancer.

2.2.2. Photodynamic Action of Resveratrol

Trans-Resveratrol has been reported to have significant photodynamic activity. The generation of singlet oxygen by *trans*-Resveratrol after light irradiation is used as PS for the photooxidation of ergosterol via a [4+2] cycloaddition (Figure 9) [99].

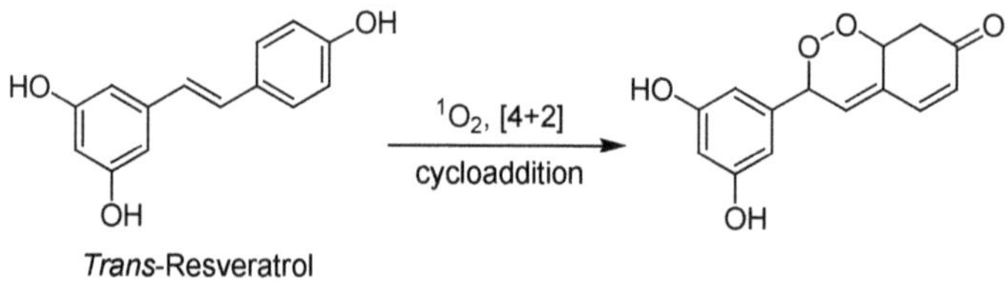

Figure 9. *Trans*-Resveratrol photooxidation of ergosterol via a [4+2] cycloaddition.

Singlet oxygen is a ROS generated by ground-state (triplet) oxygen excitation. This is the energy transfer from a photoexcited PS in the PDT process [100]. Resveratrol quinone is the main source for quenching $^1O_2$. Its mechanism is based on the resorcinol moiety and the carbon–carbon double bond [101]. It has an excellent triplet-quenching activity in a lower concentration range of 52 μM [102].

Regarding *trans*-Resveratrol with antibacterial and anticancer functions, the ROS causes the inactivation of bacteria, which oxidizes proteins or lipids, leading to bacteria death [103]. It also inhibits the proliferation of cancer cells due to increased metabolic rate, gene mutation, and relative hypoxia [104], as well as suppressing cell growth accompanied by apoptosis evoked through increased intracellular ROS levels in mitochondria [105]. Some examples of Resveratrol as a PS for PDT against bacteria and cancer are shown in Table 5.

Table 5. Photodynamic action of Resveratrol against bacteria and cancer.

		Bacteria			
	Study	**Natural Photosensitizer and Dosage**	**Usage of Light and Energy (J)**	**Consequences**	**Reference**
1	Photodynamic Therapy with Resveratrol and an Nd:YAG Laser for *Enterococcus faecalis* Elimination	357 μM of Resveratrol and Nd:YAG on *Enterococcus faecalis*.	Irradiated with red light at 635 nm in 3.5 W/cm^2, and fluence 14 J/cm^2 for 48 h.	PDT with Resveratrol and Nd:YAG as the PS with pigment was efficacious for *Enterococcus faecalis* elimination without inducing any toxic effects on osteoblasts.	[106]

Table 5. *Cont.*

	Bacteria				
	Study	**Natural Photosensitizer and Dosage**	**Usage of Light and Energy (J)**	**Consequences**	**Reference**
2	Photoactivated Resveratrol against *Staphylococcus aureus* infection in mice	No report of the Resveratrol use on *Staphylococcus aureus*	Irradiated with blue light at 450–495 nm in 75 W/cm^2, and fluence 54 J/cm^2 for 24 h.	PDT with Resveratrol generated singlet oxygen effects on the immune system, triggering TNF-α and IL-17A production to the clearance of *Staphylococcus aureus*.	[107]
3	Photoactivated Resveratrol controls intradermal infection by *Staphylococcus aureus* in mice: a pilot study	2 μM of Resveratrol on *Staphylococcus aureus*	Irradiated with blue light at 450 nm in 75 W/cm^2, and fluence 22.5 J/cm^2 for 5 min.	PDT with Resveratrol induced the expression of myeloperoxidase, greater bacterial clearance, and infection control by IL-10 production.	[108]
	Cancer				
	Study	**Natural Photosensitizer and Dosage**	**Usage of Light and Energy (J)**	**Consequences**	**Reference**
1	The comparative effects of Resveratrol and Curcumin in combination with photodynamic therapy	10 mg/kg body weight of Resveratrol and 50 mg/kg body weight of Curcumin dissolved in 0.5% of carboxymethyl cellulose through oral gavage for 7 days, respectively, on walker carcinosarcoma in 66 Wistar Male Rats.	Irradiated with laser light at 685 nm in 25 W/cm^2, and fluence 50 J/cm^2 for 15 min.	PDT with Resveratrol and Curcumin decreased oxidative stress, diminished the Cyclooxygenase-2 and Nitric oxide synthase 2 expressions, and increased cell death by positively influencing the necrotic rate and apoptotic index.	[109]
2	Resveratrol enhances the effects of ALA-PDT on skin squamous cells A431 through the p38/MAPK signaling pathway	0.5 mM of Aminolevulinic Acid as PS and 58 μM of Resveratrol is an adjuvant on Skin Squamous Cells, A431.	Irradiated with red light at 635 nm in 50 W/cm^2, and fluence 37 J/cm^2 for 3 h.	Resveratrol enhanced the effect of ALA-PDT against skin squamous cells A431 through the p38/ MAPK pathway.	[110]

3. Photodynamic Action of Hydrogel

Hydrogels are an effective way to release drugs and antibacterial agents that greatly improve the utilization of antibacterial agents in bacteria and reduce the toxic effects on normal cells [111,112]. It is also a potent novel medication that targets cancer cells and reduces damage to normal tissues during PDT treatment. They act as a powerful drug-delivery capacity for a precisely controlled drug release because hydrogels can be loaded for many therapeutic agents, such as chemotherapeutic agents, radionuclides, and immunosuppressants, to induce a cascade of multiple therapeutic modalities. This can operate as a carrier and release system of PS for PDT, which also enhances the light-penetration depth and enables an integrated sequence of cancer treatment [113].

Hydrogels have a variety of sizes, from macrogels, microgels (0.5 to 10 μM), and nanogels (less than 200 nm) during the PDT treatment, which allows precise access to the cancer site and provides continuous as well as controlled drug delivery for the selectivity of tumors. This also minimizes the usage of drugs and reduces systemic toxicity to the corresponding tissues [114]. The hydrogel system is highly dependent on internal and external environmental stimulations, such as pH, temperature, redox potential, and reactive oxygen concentration. Meanwhile, oxygen dependence is an important factor for PDT treatment. Thus, the hydrogel system can target cancer cells and release drugs to improve the therapeutic effectiveness of solid tumors [115]. Some examples of hydrogel for PDT against bacteria and cancer (Table 6).

PDT can lead to additional crosslinking within the hydrogel network, which reacts with polymer chains, altering the mechanical properties and swelling behavior of the hydrogel. The incorporation of photosensitizers into hydrogels greatly enhances their localized concentration and biocompatibility. It prolongs their residence time to generate ROS against bacteria and cancer during the PDT process [9]. However, PDT may also cause the degradation of hydrogel to release a drug (e.g., Resveratrol) since ROS breaks down polymer chains at the same time, resulting in fragmentation and affecting overall stability. The drug-delivery system is the most important point for the hydrogel applications on antibacterial and anticancer functions [116].

Table 6. Photodynamic action of hydrogel against bacteria and cancer.

		Bacteria			
	Study	**Photosensitizer and Dosage**	**Usage of Light and Energy (J)**	**Consequences**	**Reference**
1	Photosensitizer-loaded hydrogels for photodynamic inactivation of multiresistant bacteria in wounds	326 μM of tetrakis(1 methylpyridinium-4-yl)porphyrin p-toluenesulfonate (TMPyP) and 242 μM of tetrahydroporphyrin—p toluenesulfonate (THPTS) on *Enterococcus faecium*, *Staphylococcus aureus*, *Klebsiella pneumonia*, *Acinetobacter baumannii*, *Pseudomonas aeruginosa*, *Escherichia coli*, and *Achromobacter xylosoxidans*.	Irradiated with red light at 760 nm in 18 W/cm^2, and fluence 20 J/cm^2 for 36 to 90 min.	TMPyP-loaded hydrogels were more effective than those loaded with THPTS, which displayed effectivity against all investigated bacteria strains, improving the treatment of wounds infected with problematic bacterial pathogens.	[117]
2	An injectable dipeptide–fullerene supramolecular hydrogel for photodynamic antibacterial therapy	200 μM of dipeptide–fullerene supramolecular hydrogel on *Staphylococcus aureus* in wound healing.	Irradiated with white light at 400 nm in 0.1 W/cm^2, and fluence 10 J/cm^2 for 10 min.	Peptide fullerene hydrogels inhibited multi-antibiotic-resistant *Staphylococcus aureus* and promote wound healing.	[118]
3	Optimization of hydrogel containing toluidine blue O for photodynamic therapy in treating acne	0.1 μM of toluidine blue O on *Propionibacterium acnes*, *Staphylococcus aureus*, and *Escherichia coli*.	Irradiated with red light at 630 nm in 0.4 W/cm^2, and fluence 13 J/cm^2 for 15 min.	Toluidine blue O hydrogel for PDT showed effective antibacterial activity for *Propionibacterium acnes*, *Staphylococcus aureus*, and *Escherichia coli*.	[119]
		Cancer			
	Study	**Photosensitizer and Dosage**	**Usage of Light and Energy (J)**	**Consequences**	**Reference**
1	Synthesis and Characterization of Temperature-sensitive and Chemically Crosslinked Poly(N-isopropylacrylamide)/Photosensitizer Hydrogels for Applications in Photodynamic Therapy	90 μM of Pheophorbide a-poly(N-isopropylacrylamide) nanohydrogel on Human Colorectal Adenocarcinoma, HT29.	Irradiated with white light at 681 nm in 110 W/cm^2 and fluence 20 J/cm^2 for 18 to 24 h.	Pheophorbide a-poly(N-isopropylacrylamide) nanohydrogel with reasonable biocompatibility and acceptable photocytotoxicity in the low μM range.	[120]

Table 6. *Cont.*

	Cancer				
	Study	**Photosensitizer and Dosage**	**Usage of Light and Energy (J)**	**Consequences**	**Reference**
2	Alginate-Based Microcapsules with a Molecule Recognition Linker and Photosensitizer for the Combined Cancer Treatment	30 μM of Ca-ALG (HB-lipid), and Ca-ALG-DOX-(HB-lipid) hydrogels on Immortal cells, HeLa.	Irradiated with blue light at 488 nm in 0.5 W/cm^2, and fluence 30 J/cm^2 for 36 h.	Ca-ALG (HB-lipid) and Ca-ALG-DOX-(HB-lipid) hydrogels are the co-delivery carriers with high efficiency in treating PDT against Immortal cells (HeLa).	[121]
3	Curcumin and silver nanoparticles carried out from polysaccharide-based hydrogels improved the photodynamic properties of curcumin through metal-enhanced singlet oxygen effect	91.5 μM of CHT/CS/CUR-AgNPs hydrogel on Human Colon Cancer cells, Caco-2.	Irradiated with green light at 525 nm in 420 W/cm^2, and fluence 50 J/cm^2 for 24 h.	PDT selective illumination led to the inhibition of Human Colon Cancer cells (Caco-2) by the CHT/CS/CU R-AgNPs hydrogel, and CUR can work as a diagnostic fluorescence probe in this system.	[122]

3.1. *Resveratrol Combined with Hydrogel*

Resveratrol has limitations in clinical studies, including poor bioavailability, low water solubility, and chemical instability in neutral and alkaline environments [123]. Hydrogel is a special material, and Resveratrol overcomes the above issue with the help of this. A hydrogel has several properties, such as high biocompatibility, drug protection, spatiotemporal control of the drug release, and physicochemical tailorability. Some examples are summarized in Table 7.

Table 7. Hydrogel with Resveratrol against bacteria and cancers.

	Bacteria				
	Study	**Models**	**Dosage and Time**	**Consequences**	**Reference**
1	Resveratrol-Loaded Hydrogel Contact Lenses with Antioxidant and Antibiofilm Performance	*Pseudomonas aeruginosa*, and *Staphylococcus aureus*	100 and 200 μM of Resveratrol for 24 h.	Resveratrol released from the hydrogels readily accumulated in tissues and was effective against *Pseudomonas aeruginosa*, and *Staphylococcus aureus*.	[124]
2	Liposomes-In-Hydrogel Delivery System Enhances the Potential of Resveratrol in Combating Vaginal Chlamydia Infection	*Chlamydia trachomatis*	1.5 and 3 μM of Resveratrol for 48 h.	The anti-chlamydial effect of RES was enhanced when incorporated into a liposomes-in-hydrogel delivery system, which was a promising option for the localized treatment of *Chlamydia trachomatis* infection.	[125]
3	Incorporation of Resveratrol–Hydroxypropyl-β-Cyclodextrin Complexes into Hydrogel Formulation for Wound Treatment	*Staphylococcus aureus*, *Escherichia coli*, and *Candida albicans*	0.35 and 0.175 μM of Resveratrol–Hydroxypropyl-β-Cyclodextrin Complexes.	Resveratrol–Hydroxypropyl-β-Cyclodextrin Complexes were included in Pluronic hydrogel, which provided efficient drug release and appropriate viscoelastic properties for wound treatment.	[126]

Table 7. *Cont.*

	Bacteria				
	Study	**Models**	**Dosage and Time**	**Consequences**	**Reference**
4	Breathable hydrogel dressings containing natural antioxidants for the management of skin disorders	*Staphylococcus aureus*	15 and 30 μM of Resveratrol.	The developed hydrogel patch was breathable and able to maintain excellent mechanical properties with Resveratrol for 72 h against bacterial growth.	[127]
5	Resveratrol therapeutics combines both antimicrobial and immunomodulatory properties against respiratory infection by nontypeable Haemophilus influenzae	*Escherichia coli*, and *Bacillus subtilis*	112.5 and 56.25 μM of Resveratrol.	Resveratrol was therapeutic in targeting chronic obstructive pulmonary disease airway infections.	[128]
	Cancer				
	Study	**Models**	**Dosage and Time**	**Consequences**	**Reference**
1	Injectable click-crosslinked hydrogel containing Resveratrol to improve the therapeutic effect in triple-negative breast cancer	Breast cancer cells, MDA-MB-231.	0.5 μM of Resveratrol and Resveratrol with hyaluronic acid for 24 h.	Resveratrol with hyaluronic acid significantly reduced negative tumor growth rates coupled with large apoptotic cells and limited angiogenesis in tumors.	[129]
2	Chitosan-based injectable in situ forming hydrogels containing dopamine-reduced graphene oxide and Resveratrol for breast cancer chemo-photothermal therapy	Breast cancer cells, MCF-7.	66 μM of Dopamine-reduced graphene oxide with Resveratrol for 24 h.	Resveratrol-formulated hydrogels displayed injectability and in situ gelation, as well as suitable physicochemical properties and good cytocompatibility, which was an enormous potential for the chemo-photothermal therapy of breast cancer cells.	[130]
3	Thermosensitive Hydrogels Loaded with Resveratrol Nanoemulsion: Formulation Optimization by Central Composite Design and Evaluation in MCF-7 Human Breast Cancer Cell Lines.	Breast cancer cells, MCF-7.	25 μM of Resveratrol with hydrogel for 6 h.	The developed Resveratrol with hydrogel was an effective delivery of breast cancer, and the in vitro release profile demonstrated a release rate of 80%.	[131]
4	Delivery of Resveratrol, a Red Wine Polyphenol, from Solutions and Hydrogels via the Skin	Female nude mice (ca. 8 weeks old) on skin erythema.	0.6 mL aliquot of Resveratrol hydrogel for 24 h.	Resveratrol with the hydrogel caused no stratum corneum disruption or skin erythema, and it was a therapeutic skin route of administration.	[132]

Mechanism of Combination

The mechanism for the combination of Resveratrol with hydrogel against bacteria and cancer is based on the antimicrobial peptide, antimicrobial agents, antibiotics, and polysaccharide, e.g., chitosan or cyclodextrin, e.g., (i) incorporation of Resveratrol–hydroxypropyl-β-cyclodextrin complexes into hydrogel formulation for wound treatment against *Staphylococcus aureus*, *Escherichia coli*, and *Candida albicans* (Figure 10) [126], and (ii) chitosan-based injectable in situ forming hydrogels containing dopamine-reduced graphene oxide and Resveratrol for breast cancer chemo-photothermal therapy (Figure 11) [130].

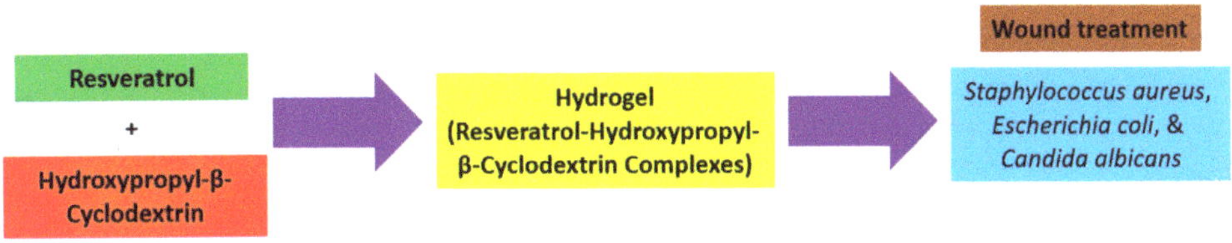

Figure 10. Synthetic diagram for the incorporation of Resveratrol–hydroxypropyl–β-cyclodextrin complexes into hydrogel formulation against bacteria.

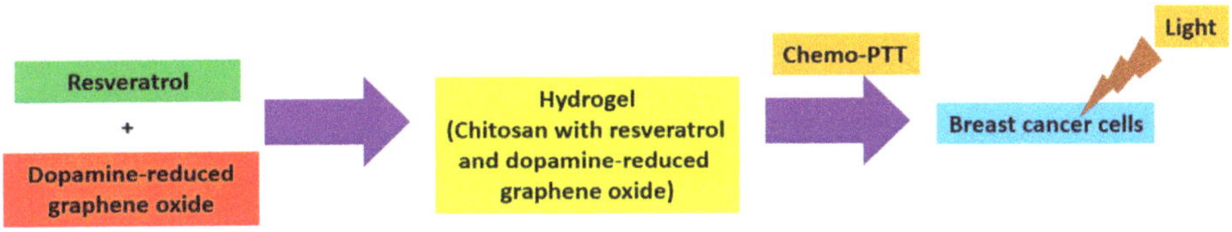

Figure 11. Synthetic diagram for the chitosan-based hydrogels containing dopamine-reduced graphene oxide and Resveratrol against breast cancer.

Based on the above information, it is possible to have a photodynamic action of Resveratrol with hydrogel against bacteria and cancer since the hydrogel is a three-dimensional (3D) network structure with crosslinked polymer chains that forms the ionic or covalent bond for linking one polymer chain to another, which provides high biocompatibility, drug protection, spatiotemporal control of the drug release, and physicochemical tailorability.

Resveratrol has antibacterial and anticancer effects. It affects the changes in cell morphology and DNA contents [133]. It is also able to inhibit the carcinogenesis stages, including initiation, promotion, and progression [134–136]. These are the reasons for using Resveratrol as a PS or adjuvant.

There is potential for the photodynamic activity of Resveratrol for treating various types of bacteria and cancers, with the particular advantage of causing minimal toxic side effects. This may increase the sensitivity of some cancer cells against chemotherapy drugs and overcome one or more of the body's mechanisms, which eliminate the side effects of chemotherapy, such as anorexia, fatigue, depression, nerve pain, and cognitive impairment [137].

Regarding the photodynamic activity of Resveratrol, it is much better combined with the hydrogel. This eliminates the clinical application problems of Resveratrol, such as poor bioavailability, low water solubility, and chemical instability in neutral and alkaline environments. It greatly enhances the effectiveness of PDT against bacteria and cancer.

Moreover, hydrogel has started to be used with nanoparticles in recent research. Yang et al. reported that Resveratrol induced apoptosis and inhibited tumor growth after being coated with gold nanoflowers. It also enriched the tumor sites and identified tumor sites by computed tomography (CT) [138]. Xiang et al. identified Resveratrol-loaded dual-function titanium disulfide nanosheets for photothermal-triggered tumor chemotherapy with no remarkable tissue toxicity. Titanium disulfide nanosheets with Resveratrol can target and accumulate in mitochondria when triggered by near-infrared light. It induces the upregulation of key intrinsic apoptotic factors such as cytochrome c and initiates the caspase cascade to achieve the chemotherapeutic effect [139]. Thus, hydrogel nanoparticles or nanocomposites should be used in further investigation of Resveratrol with PDT in the treatment of bacteria and cancer.

However, there is some toxicity, and adverse effects were reported for the consumption of Resveratrol. Therefore, extensive studies on long-term effects and the in vivo adverse effects of Resveratrol supplementation in humans are needed [140]. These include concerns about the dosage of Resveratrol and duration time for humans if it is used as a PS or adjuvant.

4. Conclusions

Resveratrol combined with hydrogel is suitable for PDT treatment to fight against bacteria and cancer. Hydrogel consists of three-dimensional (3D) network structures with natural, synthetic, or semi-synthetic polymers through physical or chemical crosslinked methods. It has several competitive advantages, including good biocompatibility and low cytotoxicity, enhanced antitumor effect, reduced toxic and side effects, as well as maintaining a high concentration and archive the selective of a drug during the PDT process.

Resveratrol is a natural polyphenol stilbene structure containing two isomeric *cis*- and *trans*-forms. *Trans*-Resveratrol is dominant and more stable with significant photodynamic activity, which acts as a PS to generate ROS during the PDT process. However, *Trans*-Resveratrol has some limitations, with poor water solubility the major issue for biological application. Thus, they are compatible and reinforce each other to increase the effectiveness of PDT against bacteria and cancer.

However, more work is required, especially for the cytotoxicity safety assessment of the human body. The selection of hydrogel and Resveratrol acting as PS or adjuvant to enhance the effectiveness of PDT is another important milestone in the future.

5. Future Aspects

How do we enhance the effectiveness and therapeutic effect of PDT with natural products against bacteria and cancer in the future? There are some strategies to improve photodynamic therapy efficacy, such as making good use of non-reactive oxygen carriers (microbubbles/nano-bubbles, hemoglobin, and perfluorocarbon), reactive oxygen carriers (PDT dependent/independent materials), and regulating the microenvironment (blood perfusion, target mitochondria, moderate the level of Hypoxia-inducible factor 1, and hypoxia-activated therapy) [141]. Nanotechnology is another useful approach including the applications of liposomes, nanoparticles, and quantum dots [142]. In 2022, Li et al. developed the nanocomposite AuNS@ZrTCPP-GA (AZG), containing gambogic acid (GA), heat-shock protein 90 (HSP90) inhibitor, and the gold nanostars (AuNS) coated with PEGylated liposome (LP) to increase the stability and biocompatibility for enhancing the anticancer effect of PDT [143].

Meanwhile, it is suggested that the combination of photodynamic (PD) and photothermal (PT) therapies harness light to eliminate cancer cells with spatiotemporal precision by either generating ROS or increasing temperature. This addresses the limitations of the PDT/PTT modality and enhances treatment safety as well as efficacy. However, the complicated preclinical assessment of PDT/PTT combinations and possible rationale or guidelines to elucidate the mechanisms underlying PDT/PTT interactions are required for further investigations [144].

Author Contributions: Conceptualization, S.K.L., C.W.S.T. and C.W.C.L.; writing—original draft preparation, S.K.L., C.W.S.T. and C.W.C.L.; writing—review and editing, S.K.L., C.W.S.T. and C.W.C.L.; supervision, S.K.L. and D.C.T.A.; funding acquisition, D.C.T.A. All authors have read and agreed to the published version of the manuscript.

Funding: This research was funded by the Chinese Medicine and Culture Research Center, Research Matching Grant (RMG030a).

Data Availability Statement: Data are contained within the article.

Conflicts of Interest: The authors declare no conflicts of interest.

References

1. Bernier, S.P.; Surette, M.G. Concentration-dependent activity of antibiotics in natural environments. *Front. Microbiol.* **2013**, *4*, 20. [CrossRef] [PubMed]
2. Friedman, N.D.; Temkin, E.; Carmeli, Y. The negative impact of antibiotic resistance. *Clin. Microbiol. Infect.* **2016**, *22*, 416–422. [CrossRef] [PubMed]
3. Grin, M.; Suvorov, N.; Ostroverkhov, P.; Pogorilyy, V.; Kirin, N.; Popov, A.; Sazonova, A.; Filonenko, E. Advantages of combined photodynamic therapy in the treatment of oncological diseases. *Biophys. Rev.* **2022**, *14*, 941–963. [CrossRef]
4. Gu, B.; Wang, B.; Li, X.; Feng, Z.; Ma, C.; Gao, L.; Yu, Y.; Zhang, J.; Zheng, P.; Wang, Y.; et al. Photodynamic therapy improves the clinical efficacy of advanced colorectal cancer and recruits immune cells into the tumor immune microenvironment. *Front. Immunol.* **2022**, *13*, 1050421. [CrossRef] [PubMed]
5. Tian, B.; Liu, J. Resveratrol: A review of plant sources, synthesis, stability, modification and food application. *J. Sci. Food Agric.* **2020**, *100*, 1392–1404. [CrossRef] [PubMed]
6. Zhao, Y.; Shi, M.; Ye, J.H.; Zheng, X.Q.; Lu, J.L.; Liang, Y.R. Photo-induced chemical reaction of trans-resveratrol. *Food Chem.* **2015**, *171*, 137–143. [CrossRef]
7. Liu, Y.; Qin, R.; Zaat, S.A.J.; Breukink, E.; Heger, M. Antibacterial photodynamic therapy: Overview of a promising approach to fight antibiotic-resistant bacterial infections. *J. Clin. Transl. Res.* **2015**, *1*, 140–167.
8. Bhattacharya, D.; Mukhopadhyay, M.; Shivam, K.; Tripathy, S.; Patra, R.; Pramanik, A. Recent developments in photodynamic therapy and its application against multidrug resistant cancers. *Biomed. Mater.* **2023**, *18*, 062005. [CrossRef]
9. Gan, S.; Wu, Y.; Zhang, X.; Zheng, Z.; Zhang, M.; Long, L.; Liao, J.; Chen, W. Recent Advances in Hydrogel-Based Phototherapy for Tumor Treatment. *Gels* **2023**, *9*, 286. [CrossRef]
10. Liu, B.; Chen, K. Advances in Hydrogel-Based Drug Delivery Systems. *Gels* **2024**, *10*, 262. [CrossRef]
11. Maitra, J.; Shukla, V.K. Cross-linking in Hydrogels—A Review. *Am. J. Polym. Sci.* **2014**, *4*, 25–31.
12. Gulati, N.; Nagaich, U.; Sharma, V.K.; Khosa, R.L. Effect of Polymer and Cross Linking Agent on In Vitro Release of Quercetin from Microbeads. *Asian J. Pharm. Sci.* **2011**, *1*, 2231–4423.
13. Denizli, B.K.; Can, H.K.; Rzaev, Z.M.O.; Guner, A. Preparation conditions and swelling equilibria of dextran hydrogels prepared by some crosslinked agents. *Polymer* **2004**, *45*, 6431–6435. [CrossRef]
14. Zhu, J. Bioactive modification of poly(ethylene glycol) hydrogels for tissue engineering. *Biomaterials* **2010**, *31*, 4639–4656. [CrossRef]
15. Ramamurthi, A.; Vesely, I. Ultraviolet light-induced modification of crosslinked hyaluronan gels. *J. Biomed. Mater. Res. A* **2003**, *66*, 317–329. [CrossRef]
16. Ho, T.C.; Chang, C.C.; Chan, H.P.; Chung, T.W.; Shu, C.W.; Chuang, K.P.; Duh, T.H.; Yang, M.H.; Tyan, Y.C. Hydrogels: Properties and Applications in Biomedicine. *Molecules* **2022**, *27*, 2902. [CrossRef]
17. Foudazi, R.; Zowada, R.; Manas-Zloczower, I.; Feke, D.L. Porous Hydrogels: Present Challenges and Future Opportunities. *Langmuir* **2023**, *39*, 2092–2111. [CrossRef]
18. Bouklas, N.; Huang, R. Swelling kinetics of polymer gels: Comparison of linear and nonlinear theories. *Soft Matter* **2012**, *8*, 8194–8203. [CrossRef]
19. Yan, H.; Jin, B. Equilibrium swelling of a polyampholytic pH-sensitive hydrogel. *Eur. Phys. J. E Soft Matter* **2013**, *36*, 27. [CrossRef]
20. Xu, Z.; Chen, Y.; Cao, Y.; Xue, B. Tough Hydrogels with Different Toughening Mechanisms and Applications. *Int. J. Mol. Sci.* **2024**, *25*, 2675. [CrossRef]
21. Axpe, E.; Chan, D.; Offeddu, G.S.; Chang, Y.; Merida, D.; Hernandez, H.L.; Appel, E.A. A Multiscale Model for Solute Diffusion in Hydrogels. *Macromolecules* **2019**, *52*, 6889–6897. [CrossRef] [PubMed]
22. Fasiku, V.O.; Omolo, C.A.; Devnarain, N.; Ibrahim, U.H.; Rambharose, S.; Faya, M.; Mocktar, C.; Singh, S.D.; Govender, T. Chitosan-Based Hydrogel for the Dual Delivery of Antimicrobial Agents against Bacterial Methicillin-Resistant *Staphylococcus aureus* Biofilm-Infected Wounds. *ACS Omega* **2021**, *6*, 21994–22010. [CrossRef] [PubMed]
23. Abbasalizadeh, F.; Alizadeh, E.; Bagher Fazljou, S.M.; Torbati, M.; Akbarzadeh, A. Anticancer Effect of Alginate-chitosan Hydrogel Loaded with Curcumin and Chrysin on Lung and Breast Cancer Cell Lines. *Curr. Drug Deliv.* **2022**, *19*, 600–613. [PubMed]
24. Lu, H.; Huang, Y.; Lv, F.; Liu, L.; Ma, Y.; Wang, S. Living Bacteria-Mediated Aerobic Photoinduced Radical Polymerization for in Situ Bacterial Encapsulation and Differentiation. *CCS Chem.* **2021**, *3*, 1296–1305. [CrossRef]
25. Andrade, F.; Roca-Melendres, M.M.; Durán-Lara, E.F.; Rafael, D.; Schwartz, S., Jr. Stimuli-Responsive Hydrogels for Cancer Treatment: The Role of pH, Light, Ionic Strength and Magnetic Field. *Cancers* **2021**, *13*, 1164. [CrossRef]
26. Nairon, K.G.; DePalma, T.; Sivakumar, H.; Skardal, A. Tunable Hydrogel Systems for Delivery and Release of Cell-Secreted and Synthetic Therapeutic Products. In *Controlled Drug Delivery Systems*, 1st ed.; CRC Press: Boca Raton, FL, USA, 2020; Volume 3, pp. 29–53.
27. Naahidi, S.; Jafari, M.; Logan, M.; Wang, Y.; Yuan, Y.; Bae, H.; Dixon, B.; Chen, P. Biocompatibility of hydrogel-based scaffolds for tissue engineering applications. *Biotechnol. Adv.* **2017**, *35*, 530–544. [CrossRef]
28. Li, J.; Mooney, D.J. Designing hydrogels for controlled drug delivery. *Nat. Rev. Mater.* **2016**, *1*, 16071. [CrossRef]
29. Rahoui, N.; Jiang, B.; Taloub, N.; Huang, Y.D. Spatio-temporal control strategy of drug delivery systems based nano structures. *J. Control. Release* **2017**, *255*, 176–201. [CrossRef]

30. Klotz, B.J.; Gawlitta, D.; Rosenberg, A.J.W.P.; Malda, J.; Melchels, F.P.W. Gelatin-Methacryloyl Hydrogels: Towards Biofabrication-Based Tissue Repair. *Trends Biotechnol.* **2016**, *34*, 394–407. [CrossRef]
31. Chang, G.; Zhang, H.; Li, S.; Huang, F.; Shen, Y.; Xie, A. Effective photodynamic therapy of polymer hydrogel on tumor cells prepared using methylene blue sensitized mesoporous titania nanocrystal. *Mater. Sci. Eng. C Mater. Biol. Appl.* **2019**, *99*, 1392–1398. [CrossRef]
32. Correia, J.H.; Rodrigues, J.A.; Pimenta, S.; Dong, T.; Yang, Z. Photodynamic Therapy Review: Principles, Photosensitizers, Applications, and Future Directions. *Pharmaceutics* **2021**, *13*, 1332. [CrossRef] [PubMed]
33. Ji, B.; Wei, M.; Yang, B. Recent advances in nanomedicines for photodynamic therapy (PDT)-driven cancer immunotherapy. *Theranostics* **2022**, *12*, 434–458. [CrossRef] [PubMed]
34. Hu, X.; Zhang, H.; Wang, Y.; Shiu, B.C.; Lin, J.H.; Zhang, S.; Lou, C.W.; Li, T.T. Synergistic antibacterial strategy based on photodynamic therapy: Progress and perspectives. *Chem. Eng. J.* **2022**, *3*, 138129. [CrossRef]
35. Jiang, W.; Liang, M.; Lei, Q.; Li, G.; Wu, S. The Current Status of Photodynamic Therapy in Cancer Treatment. *Cancers* **2023**, *15*, 585. [CrossRef]
36. Jiang, J.; Lv, X.; Cheng, H.; Yang, D.; Xu, W.; Hu, Y.; Song, Y.; Zeng, G. Type I photodynamic antimicrobial therapy: Principles, progress, and future perspectives. *Acta Biomater.* **2024**, *177*, 1–19. [CrossRef]
37. Yanten, N.; Vilches, S.; Palavecino, C.E. Photodynamic therapy for the treatment of *Pseudomonas aeruginosa* infections: A scoping review. *Photodiagnosis Photodyn. Ther.* **2023**, *44*, 103803. [CrossRef]
38. Gourlot, C.; Gosset, A.; Glattard, E.; Aisenbrey, C.; Rangasamy, S.; Rabineau, M.; Ouk, T.S.; Sol, V.; Lavalle, P.; Gourlaouen, C.; et al. Antibacterial Photodynamic Therapy in the Near-Infrared Region with a Targeting Antimicrobial Peptide Connected to a π-Extended Porphyrin. *ACS Infect. Dis.* **2022**, *8*, 1509–1520. [CrossRef]
39. He, L.; Yu, X.; Li, W. Recent Progress and Trends in X-ray-Induced Photodynamic Therapy with Low Radiation Doses. *ACS Nano* **2022**, *16*, 19691–19721. [CrossRef]
40. Alsaab, H.O.; Alghamdi, M.S.; Alotaibi, A.S.; Alzhrani, R.; Alwuthaynani, F.; Althobaiti, Y.S.; Almalki, A.H.; Sau, S.; Iyer, A.K. Progress in Clinical Trials of Photodynamic Therapy for Solid Tumors and the Role of Nanomedicine. *Cancers* **2020**, *12*, 2793. [CrossRef]
41. Penetra, M.; Arnaut, L.G.; Gomes-da-Silva, L.C. Trial watch: An update of clinical advances in photodynamic therapy and its immunoadjuvant properties for cancer treatment. *Oncoimmunology* **2023**, *12*, 2226535. [CrossRef]
42. Huang, L.; Xuan, Y.; Koide, Y.; Zhiyentayev, T.; Tanaka, M.; Hamblin, M.R. Type I and Type II mechanisms of antimicrobial photodynamic therapy: An in vitro study on gram-negative and gram-positive bacteria. *Lasers Surg. Med.* **2012**, *44*, 490–499. [CrossRef] [PubMed]
43. Foote, C.S. Definition of type I and type II photosensitized oxidation. *Photochem. Photobiol.* **1991**, *54*, 659. [CrossRef] [PubMed]
44. Ghorbani, J.; Rahban, D.; Aghamiri, S.; Teymouri, A.; Bahador, A. Photosensitizers in antibacterial photodynamic therapy: An overview. *Laser Ther.* **2018**, *27*, 293–302. [CrossRef] [PubMed]
45. Huang, K.; Wang, J.; Zhuang, A.; Liu, Q.; Li, F.; Yuan, K.; Yang, Y.; Liu, Y.; Chang, H.; Liang, Y.; et al. Metallacage-based enhanced PDT strategy for bacterial elimination via inhibiting endogenous NO production. *Proc. Natl. Acad. Sci. USA* **2023**, *120*, e2218973120. [CrossRef] [PubMed]
46. Mishchenko, T.; Balalaeva, I.; Gorokhova, A.; Vedunova, M.; Dmitri, V. Krysko. Which cell death modality wins the contest for photodynamic therapy of cancer? *Cell Death Dis.* **2022**, *13*, 455. [CrossRef] [PubMed]
47. Kwiatkowski, S.; Knap, B.; Przystupski, D.; Saczko, J.; Kędzierska, E.; Knap-Czop, K.; Kotlińska, J.; Michel, O.; Kotowski, K.; Kulbacka, J. Photodynamic therapy—Mechanisms, photosensitizers and combinations. *Biomed. Pharmacother.* **2018**, *106*, 1098–1107. [CrossRef]
48. Tang, P.M.; Bui-Xuan, N.H.; Wong, C.K.; Fong, W.P.; Fung, K.P. Pheophorbide a-Mediated Photodynamic Therapy Triggers HLA Class I-Restricted Antigen Presentation in Human Hepatocellular Carcinoma. *Transl. Oncol.* **2010**, *3*, 114–122. [CrossRef]
49. Hu, Y.; Zhang, C.; Li, S.; Jiao, Y.; Qi, T.; Wei, G.; Han, G. Effects of Photodynamic Therapy Using Yellow LED-light with Concomitant Hypocrellin B on Apoptotic Signaling in Keloid Fibroblasts. *Int. J. Biol. Sci.* **2017**, *13*, 319–326. [CrossRef]
50. Kah, G.; Chandran, R.; Abrahamse, H. Curcumin a Natural Phenol and Its Therapeutic Role in Cancer and Photodynamic Therapy: A Review. *Pharmaceutics* **2023**, *15*, 639. [CrossRef]
51. Sulaiman, C.; George, B.P.; Balachandran, I.; Abrahamse, H. Photoactive Herbal Compounds: A Green Approach to Photodynamic Therapy. *Molecules* **2022**, *27*, 5084. [CrossRef]
52. Harvey, A.L.; Edrada-Ebel, R.; Quinn, R.J. The Re-Emergence of Natural Products for Drug Discovery in the Genomics Era. *Nat. Rev. Drug Discov.* **2015**, *14*, 111–129. [CrossRef] [PubMed]
53. Kubrak, T.P.; Kołodziej, P.; Sawicki, J.; Mazur, A.; Koziorowska, K.; Aebisher, D. Some Natural Photosensitizers and Their Medicinal Properties for Use in Photodynamic Therapy. *Molecules* **2022**, *27*, 1192. [CrossRef] [PubMed]
54. Yamamoto, T.; Iriuchishima, T.; Aizawa, S.; Okano, T.; Goto, B.; Nagai, Y.; Horaguchi, T.; Ryu, J.; Saito, A. Bactericidal effect of photodynamic therapy using Na-pheophorbide a: Evaluation of adequate light source. *Photomed. Laser Surg.* **2009**, *27*, 849–853. [CrossRef] [PubMed]
55. Otieno, W.; Liu, C.; Deng, H.; Li, J.; Zeng, X.; Ji, Y. Hypocrellin B-Mediated Photodynamic Inactivation of Gram-Positive Antibiotic-Resistant Bacteria: An In Vitro Study. *Photobiomodulation Photomed. Laser Surg.* **2020**, *38*, 36–42. [CrossRef]

56. Ribeiro, I.P.; Pinto, J.G.; Souza, B.M.N.; Miñán, A.G.; Ferreira-Strixino, J. Antimicrobial photodynamic therapy with curcumin on methicillin-resistant Staphylococcus aureus biofilm. *Photodiagnosis Photodyn. Ther.* **2022**, *37*, 102729. [CrossRef]
57. Jiang, Y.; Leung, A.W.; Wang, X.; Zhang, H.; Xu, C. Effect of photodynamic therapy with hypocrellin B on apoptosis, adhesion, and migration of cancer cells. *Int. J. Radiat. Biol.* **2014**, *90*, 575–579. [CrossRef]
58. Ravera, S.; Pasquale, C.; Panfoli, I.; Bozzo, M.; Agas, D.; Bruno, S.; Hamblin, M.R.; Amaroli, A. Assessing the Effects of Curcumin and 450 nm Photodynamic Therapy on Oxidative Metabolism and Cell Cycle in Head and Neck Squamous Cell Carcinoma: An In Vitro Study. *Cancers* **2024**, *16*, 1642. [CrossRef]
59. Takaoka, M.J. Of the phenolic substances of white hellebore (*Veratrum grandiflorum* Loes. *fil.) J. Fac. Sci. Hokkaido Imp. Univ.* **1940**, *3*, 1–16. [CrossRef]
60. Paul, B.; Masih, I.; Deopujari, J.; Charpentier, C. Occurrence of resveratrol and pterostilbene in age-old darakchasava, an ayurvedic medicine from India. *J. Ethnopharmacol.* **1999**, *68*, 71–76. [CrossRef]
61. Elíes-Gómez, J. Efectos de los Isómeros del Resveratrol Sobre la Homeostasis del Calcio y del Óxido Nítrico en Células Vasculares. Ph.D. Thesis, Universidade de Santiago de Compostela, Santiago de Compostela, Spain, 2009.
62. Soleas, G.J.; Diamandis, E.P.; Goldberg, D.M. Resveratrol: A molecule whose time has come? And gone? *Clin. Biochem.* **1997**, *30*, 91–113. [CrossRef]
63. Sato, M.; Maulik, G.; Bagchi, D.; Das, D.K. Myocardial protection by protykin, a novel extract of trans-resveratrol and emodin. *Free. Radic. Res.* **2000**, *32*, 135–144. [CrossRef] [PubMed]
64. Palomino, O.; Gómez-Serranillos, M.P.; Slowing, K.; Carretero, E.; Villar, A. Study of polyphenols in grape berries by reversed-phase high-performance liquid chromatography. *J. Chromatogr. A* **2000**, *870*, 449–451. [CrossRef] [PubMed]
65. Burns, J.; Yokota, T.; Ashihara, H.; Lean, M.E.J.; Crozier, A. Plant foods and herbal sources of resveratrol. *J. Agric. Food Chem.* **2002**, *50*, 3337–3340. [CrossRef] [PubMed]
66. Gao, X.; Xu, Y.X.; Janakiraman, N.; Chapman, R.A.; Gautam, S.C. Immunomodulatory activity of resveratrol: Suppression of lymphocyte proliferation, development of cell-mediated cytotoxicity, and cytokine production. *Biochem. Pharmacol.* **2001**, *62*, 1299–1308. [CrossRef] [PubMed]
67. Mahjabeen, W.; Khan, D.A.; Mirza, S.A. Role of resveratrol supplementation in regulation of glucose hemostasis, inflammation and oxidative stress in patients with diabetes mellitus type 2: A randomized, placebo-controlled trial. *Complement. Ther. Med.* **2022**, *66*, 102819. [CrossRef]
68. Gu, J.; Li, Z.; Chen, H.; Xu, X.; Li, Y.; Gui, Y. Neuroprotective Effect of Trans-Resveratrol in Mild to Moderate Alzheimer Disease: A Randomized, Double-Blind Trial. *Neurol. Ther.* **2021**, *10*, 905–917. [CrossRef]
69. Bradamante, S.; Barenghi, L.; Villa, A. Cardiovascular protective effects of resveratrol. *Cardiovasc. Drug Rev.* **2004**, *22*, 169–188. [CrossRef]
70. de la Lastra, C.A.; Villegas, I. Resveratrol as an antioxidant and pro-oxidant agent: Mechanisms and clinical implications. *Biochem. Soc. Trans.* **2007**, *35*, 1156–1160. [CrossRef]
71. Meng, T.; Xiao, D.; Muhammed, A.; Deng, J.; Chen, L.; He, J. Anti-Inflammatory Action and Mechanisms of Resveratrol. *Molecules* **2021**, *26*, 229. [CrossRef]
72. Hwang, D.; Lim, Y.H. Resveratrol antibacterial activity against Escherichia coli is mediated by Z-ring formation inhibition via suppression of FtsZ expression. *Sci. Rep.* **2015**, *5*, 10029. [CrossRef]
73. Cebrián, R.; Li, Q.; Peñalver, P.; Belmonte-Reche, E.; Andrés-Bilbao, M.; Lucas, R.; de Paz, M.V.; Kuipers, O.P.; Morales, J.C. Chemically Tuning Resveratrol for the Effective Killing of Gram-Positive Pathogens. *J. Nat. Prod.* **2022**, *85*, 1459–1473. [CrossRef] [PubMed]
74. Karameşe, M.; Dicle, Y. The Antibacterial and Antibiofilm Activities of Resveratrol on Gram-positive and Gram-negative Bacteria. *Kafkas J. Med. Sci.* **2022**, *12*, 201–206. [CrossRef]
75. Liu, L.; Yu, J.; Shen, X.; Cao, X.; Zhan, Q.; Guo, Y.; Yu, F. Resveratrol enhances the antimicrobial effect of polymyxin B on Klebsiella pneumoniae and Escherichia coli isolates with polymyxin B resistance. *BMC Microbiol.* **2020**, *20*, 306. [CrossRef] [PubMed]
76. Migliaccio, A.; Stabile, M.; Bagattini, M.; Triassi, M.; Berisio, R.; De Gregorio, E.; Zarrilli, R. Resveratrol Reverts Tolerance and Restores Susceptibility to Chlorhexidine and Benzalkonium in Gram-Negative Bacteria, Gram-Positive Bacteria and Yeasts. *Antibiotics* **2022**, *11*, 961. [CrossRef] [PubMed]
77. Al Azzaz, J.; Rieu, A.; Aires, V.; Delmas, D.; Chluba, J.; Winckler, P.; Bringer, M.A.; Lamarche, J.; Vervandier-Fasseur, D.; Dalle, F.; et al. Resveratrol-Induced Xenophagy Promotes Intracellular Bacteria Clearance in Intestinal Epithelial Cells and Macrophages. *Front. Immunol.* **2019**, *9*, 3149. [CrossRef]
78. Zhou, M.; Niu, H.; Cui, D.; Huang, G.; Li, J.; Tian, H.; Xu, X.; Liang, F.; Chen, R. Resveratrol triggers autophagy-related apoptosis to inhibit the progression of colorectal cancer via inhibition of FOXQ1. *Phytother. Res.* **2024**, *38*, 3218–3239. [CrossRef]
79. Lee, S.H.; Lee, Y.J. Synergistic anticancer activity of resveratrol in combination with docetaxel in prostate carcinoma cells. *Nutr. Res. Pract.* **2021**, *15*, 12–25. [CrossRef]
80. Fatehi, R.; Rashedinia, M.; Akbarizadeh, A.R.; Zamani, M.; Firouzabadi, N. Metformin enhances anti-cancer properties of resveratrol in MCF-7 breast cancer cells via induction of apoptosis, autophagy and alteration in cell cycle distribution. *Biochem. Biophys. Res. Commun.* **2023**, *644*, 130–139. [CrossRef]
81. Ma, L.; Li, W.; Wang, R.; Nan, Y.; Wang, Q.; Liu, W.; Jin, F. Resveratrol enhanced anticancer effects of cisplatin on non-small cell lung cancer cell lines by inducing mitochondrial dysfunction and cell apoptosis. *Int. J. Oncol.* **2015**, *47*, 1460–1468. [CrossRef]

82. Ivanova, D.; Zhelev, Z.; Semkova, S.; Aoki, I.; Bakalova, R. Resveratrol Modulates the Redox-status and Cytotoxicity of Anticancer Drugs by Sensitizing Leukemic Lymphocytes and Protecting Normal Lymphocytes. *Anticancer. Res.* **2019**, *39*, 3745–3755. [CrossRef]
83. ALkharashi, N.A. Efficacy of resveratrol against breast cancer and hepatocellular carcinoma cell lines. *Saudi Med. J.* **2023**, *44*, 246–252. [CrossRef] [PubMed]
84. Can, G.; Cakir, Z.; Kartal, M.; Gunduz, U.; Baran, Y. Apoptotic effects of resveratrol, a grape polyphenol, on imatinib-sensitive and resistant K562 chronic myeloid leukemia cells. *Anticancer. Res.* **2012**, *32*, 2673–2678. [PubMed]
85. Du, Q.; Hu, B.; An, H.; Shen, K.; Xu, L.; Deng, S.; Wei, M. Synergistic anticancer effects of curcumin and resveratrol in Hepa1-6 hepatocellular carcinoma cells. *Oncol. Rep.* **2013**, *29*, 1851–1858. [CrossRef] [PubMed]
86. Lee, Y.; Shin, H.; Kim, J. In vivo Anti-Cancer Effects of Resveratrol Mediated by NK Cell Activation. *J. Innate Immun.* **2021**, *13*, 94–106. [CrossRef]
87. Moreira, H.; Szyjka, A.; Grzesik, J.; Pelc, K.; Żuk, M.; Kulma, A.; Emhemmed, F.; Muller, C.D.; Gąsiorowski, K.; Barg, E. Celastrol and Resveratrol Modulate SIRT Genes Expression and Exert Anticancer Activity in Colon Cancer Cells and Cancer Stem-like Cells. *Cancers* **2022**, *14*, 1372. [CrossRef]
88. Subramanian, M.; Soundar, S.; Mangoli, S. DNA damage is a late event in resveratrol-mediated inhibition of *Escherichia coli*. *Free Radic. Res.* **2016**, *50*, 708–719. [CrossRef]
89. Haranahalli, K.; Tong, S.; Ojima, I. Recent advances in the discovery and development of antibacterial agents targeting the cell-division protein FtsZ. *Bioorg. Med. Chem.* **2016**, *24*, 6354–6369. [CrossRef]
90. Subramanian, M.; Goswami, M.; Chakraborty, S.; Jawali, N. Resveratrol induced inhibition of Escherichia coli proceeds via membrane oxidation and independent of diffusible reactive oxygen species generation. *Redox Biol.* **2014**, *2*, 865–872. [CrossRef]
91. Dadi, P.K.; Ahmad, M.; Ahmad, Z. Inhibition of ATPase activity of Escherichia coli ATP synthase by polyphenols. *Int. J. Biol. Macromol.* **2009**, *45*, 72–79. [CrossRef]
92. Beall, B.; Lutkenhaus, J. FtsZ in Bacillus subtilis is required for vegetative septation and for asymmetric septation during sporulation. *Genes. Dev.* **1991**, *5*, 447–455. [CrossRef]
93. Kursvietiene, L.; Kopustinskiene, D.M.; Staneviciene, I.; Mongirdiene, A.; Kubová, K.; Masteikova, R.; Bernatoniene, J. Anti-Cancer Properties of Resveratrol: A Focus on Its Impact on Mitochondrial Functions. *Antioxidants* **2023**, *12*, 2056. [CrossRef] [PubMed]
94. Danial, N.N.; Korsmeyer, S.J. Cell death: Critical control points. *Cell* **2004**, *116*, 205–219. [CrossRef] [PubMed]
95. Van Opdenbosch, N.; Lamkanfi, M. Caspases in cell death, inflammation, and disease. *Immunity* **2019**, *50*, 1352–1364. [CrossRef] [PubMed]
96. Pistritto, G.; Trisciuoglio, D.; Ceci, C.; Garufi, A.; D'Orazi, G. Apoptosis as anticancer mechanism: Function and dysfunction of its modulators and targeted therapeutic strategies. *Aging* **2016**, *8*, 603–619. [CrossRef]
97. Degterev, A.; Boyce, M.; Yuan, J. A decade of caspases. *Oncogene* **2003**, *22*, 8543–8567. [CrossRef]
98. Kim, T.H.; Park, J.H.; Woo, J.S. Resveratrol induces cell death through ROS-dependent downregulation of Notch1/PTEN/Akt signaling in ovarian cancer cells. *Mol. Med. Rep.* **2019**, *19*, 3353–3360. [CrossRef]
99. Lagunes, I.; Trigos, Á. Photo-oxidation of ergosterol: Indirect detection of antioxidants photosensitizers or quenchers of singlet oxygen. *J. Photochem. Photobiol. B Biol.* **2015**, *145*, 30–34. [CrossRef]
100. Monsour, C.G.; Tadle, A.C.; Tafolla-Aguirre, B.J.; Lakshmanan, N.; Yoon, J.H.; Sabio, R.B.; Selke, M. Singlet Oxygen Quenching by Resveratrol Derivatives. *Photochem. Photobiol.* **2023**, *99*, 672–679. [CrossRef]
101. Jiang, L.Y.; He, S.; Jiang, K.Z.; Sun, C.R.; Pan, Y.J. Resveratrol and its oligomers from wine grapes are selective $^1O_2$ quenchers: Mechanistic implication by high-performance liquid chromatography-electrospray ionization-tandem mass spectrometry and theoretical calculation. *J. Agric. Food Chem.* **2010**, *58*, 9020–9027. [CrossRef]
102. Angelé-Martínez, C.; Goncalves, L.C.P.; Premi, S.; Augusto, F.A.; Palmatier, M.A.; Amar, S.K.; Brash, D.E. Triplet-Energy Quenching Functions of Antioxidant Molecules. *Antioxidants* **2022**, *11*, 357. [CrossRef]
103. Maisch, T.; Baier, J.; Franz, B.; Maier, M.; Landthaler, M.; Szeimies, R.M.; Bäumler, W. The role of singlet oxygen and oxygen concentration in photodynamic inactivation of bacteria. *Proc. Natl. Acad. Sci. USA* **2007**, *104*, 7223–7228. [CrossRef] [PubMed]
104. Perillo, B.; Di Donato, M.; Pezone, A.; Di Zazzo, E.; Giovannelli, P.; Galasso, G.; Castoria, G.; Migliaccio, A. ROS in cancer therapy: The bright side of the moon. *Exp. Mol. Med.* **2020**, *52*, 192–203. [CrossRef] [PubMed]
105. Arihara, Y.; Takada, K.; Kamihara, Y.; Hayasaka, N.; Nakamura, H.; Murase, K.; Ikeda, H.; Iyama, S.; Sato, T.; Miyanishi, K.; et al. Small molecule CP-31398 induces reactive oxygen species-dependent apoptosis in human multiple myeloma. *Oncotarget* **2017**, *8*, 65889–65899. [CrossRef] [PubMed]
106. Morikawa, M.; Uehara, S.; Yoshida, A.; Sakagami, H.; Masuda, Y. Photodynamic Therapy with Resveratrol and an Nd:YAG Laser for *Enterococcus faecalis* Elimination. *In Vivo* **2024**, *38*, 559–566. [CrossRef] [PubMed]
107. Dos Santos, D.P.; Soares Lopes, D.P.; de Moraes, R.C., Jr.; Gonçalves, C.V.; Rosa, L.P.; da Silva Rosa, F.C.; da Silva, R.A.A. Photoactivated resveratrol against Staphylococcus aureus infection in mice. *Photodiagnosis Photodyn. Ther.* **2019**, *25*, 227–236. [CrossRef]
108. Tosato, M.G.; Schilardi, P.L.; de Mele, M.F.L.; Thomas, A.H.; Miñán, A.; Lorente, C. Resveratrol enhancement staphylococcus aureus survival under levofloxacin and photodynamic treatments. *Int. J. Antimicrob. Agents* **2018**, *51*, 255–259. [CrossRef]

109. Laszló, I.P.; Laszló, M.R.; Popescu, T.; Toma, V.; Ion, R.M.; Moldovan, R.; Filip, G.A.; Cainap, C.; Clichici, S.; Muresan, A. The comparative effects of Resveratrol and Curcumin in combination with photodynamic therapy. *Med. Pharm. Rep.* **2022**, *95*, 165–178. [CrossRef]
110. Zhang, X.; Liu, X.; Kang, S.; Liu, C.; Hao, Y. Resveratrol enhances the effects of ALA-PDT on skin squamous cells A431 through p38/MAPK signaling pathway. *Cancer Biomark.* **2018**, *21*, 797–803. [CrossRef]
111. Hu, J.; Quan, Y.; Lai, Y.; Zheng, Z.; Hu, Z.; Wang, X.; Dai, T.; Zhang, Q.; Cheng, Y. A smart aminoglycoside hydrogel with tunable gel degradation, on-demand drug release, and high antibacterial activity. *J. Control Release* **2017**, *247*, 145–152. [CrossRef]
112. Mensah, A.; Rodgers, A.M.; Larrañeta, E.; McMullan, L.; Tambuwala, M.; Callan, J.F.; Courtenay, A.J. Treatment of periodontal infections, the possible role of hydrogels as antibiotic drug-delivery systems. *Antibiotics* **2023**, *12*, 1073. [CrossRef]
113. Su, J.; Lu, S.; Jiang, S.; Li, B.; Liu, B.; Sun, Q.; Li, J.; Wang, F.; Wei, Y. Engineered Protein Photo-Thermal Hydrogels for Outstanding In Situ Tongue Cancer Therapy. *Adv. Mater.* **2021**, *33*, e2100619. [CrossRef] [PubMed]
114. Kass, L.E.; Nguyen, J. Nanocarrier-hydrogel composite delivery systems for precision drug release. *Wiley Interdiscip. Rev. Nanomed. Nanobiotechnol.* **2022**, *14*, e1756. [CrossRef] [PubMed]
115. Jiang, Y.; Wang, Y.; Li, Q.; Yu, C.; Chu, W. Natural polymer-based stimuli-responsive hydrogels. *Curr. Med. Chem.* **2020**, *27*, 2631–2657. [CrossRef]
116. Zhang, Y.; Tian, S.; Huang, L.; Li, Y.; Lu, Y.; Li, H.; Chen, G.; Meng, F.; Liu, G.L.; Yang, X.; et al. Reactive oxygen species-responsive and Raman-traceable hydrogel combining photodynamic and immune therapy for postsurgical cancer treatment. *Nat. Commun.* **2022**, *13*, 4553. [CrossRef] [PubMed]
117. Glass, S.; Kühnert, M.; Lippmann, N.; Zimmer, J.; Werdehausen, R.; Abel, B.; Eulenburg, V.; Schulze, A. Photosensitizer-loaded hydrogels for photodynamic inactivation of multirestistant bacteria in wounds. *RSC Adv.* **2021**, *11*, 7600–7609. [CrossRef] [PubMed]
118. Zhang, Y.; Zhang, H.; Zou, Q.; Xing, R.; Jiao, T.; Yan, X. An injectable dipeptide-fullerene supramolecular hydrogel for photodynamic antibacterial therapy. *J. Mater. Chem. B* **2018**, *6*, 7335–7342. [CrossRef]
119. Zheng, Y.; Yu, E.; Weng, Q.; Zhou, L.; Li, Q. Optimization of hydrogel containing toluidine blue O for photodynamic therapy in treating acne. *Lasers Med. Sci.* **2019**, *34*, 1535–1545. [CrossRef]
120. Belali, S.; Savoie, H.; O'Brien, J.M.; Cafolla, A.A.; O'Connell, B.; Karimi, A.R.; Boyle, R.W.; Senge, M.O. Synthesis and Characterization of Temperature-Sensitive and Chemically Cross-Linked Poly(N-isopropylacrylamide)/Photosensitizer Hydrogels for Applications in Photodynamic Therapy. *Biomacromolecules* **2018**, *19*, 1592–1601. [CrossRef]
121. Du, C.; Zhao, J.; Fei, J.; Gao, L.; Cui, W.; Yang, Y.; Li, J. Alginate-Based Microcapsules with a Molecule Recognition Linker and Photosensitizer for the Combined Cancer Treatment. *Chem. Asian J.* **2013**, *8*, 736–742. [CrossRef]
122. de Freitas, C.F.; Kimura, E.; Rubira, A.F.; Muniz, E.C. Curcumin and silver nanoparticles carried out from polysaccharide-based hydrogels improved the photodynamic properties of curcumin through metal-enhanced singlet oxygen effect. *Mater. Sci. Eng. C* **2020**, *112*, 110853. [CrossRef]
123. Lorenzo, D.; Giorgio, C.; Giulia, B.; Francesco Paolo, B.; Giulio, D.; Alessandra, G.; Cristiana, I.; Matteo, N.; Enrico, T.; Chiara, T.; et al. Certainty and uncertainty in the biological activities of resveratrol. *Food Front.* **2024**, *5*, 849–854.
124. Vivero-Lopez, M.; Muras, A.; Silva, D.; Serro, A.P.; Otero, A.; Concheiro, A.; Alvarez-Lorenzo, C. Resveratrol-Loaded Hydrogel Contact Lenses with Antioxidant and Antibiofilm Performance. *Pharmaceutics* **2021**, *13*, 532. [CrossRef] [PubMed]
125. Jøraholmen, M.W.; Johannessen, M.; Gravningen, K.; Puolakkainen, M.; Acharya, G.; Basnet, P.; Škalko-Basnet, N. Liposomes-In-Hydrogel Delivery System Enhances the Potential of Resveratrol in Combating Vaginal Chlamydia Infection. *Pharmaceutics* **2020**, *12*, 1203. [CrossRef] [PubMed]
126. Radeva, L.; Yordanov, Y.; Spassova, I.; Kovacheva, D.; Tibi, I.P.; Zaharieva, M.M.; Kaleva, M.; Najdenski, H.; Petrov, P.D.; Tzankova, V.; et al. Incorporation of Resveratrol-Hydroxypropyl-β-Cyclodextrin Complexes into Hydrogel Formulation for Wound Treatment. *Gels* **2024**, *10*, 346. [CrossRef] [PubMed]
127. Comotto, M.; Saghazadeh, S.; Bagherifard, S.; Aliakbarian, B.; Kazemzadeh-Narbat, M.; Sharifi, F.; Mousavi Shaegh, S.A.; Arab-Tehrany, E.; Annabi, N.; Perego, P.; et al. Breathable hydrogel dressings containing natural antioxidants for management of skin disorders. *J. Biomater. Appl.* **2019**, *33*, 1265–1276. [CrossRef]
128. Euba, B.; López-López, N.; Rodríguez-Arce, I.; Fernández-Calvet, A.; Barberán, M.; Caturla, N.; Martí, S.; Díez-Martínez, R.; Garmendia, J. Resveratrol therapeutics combines both antimicrobial and immunomodulatory properties against respiratory infection by nontypeable Haemophilus influenzae. *Sci. Rep.* **2017**, *7*, 12860. [CrossRef]
129. Shin, G.R.; Kim, H.E.; Ju, H.J.; Kim, J.H.; Choi, S.; Choi, H.S.; Kim, M.S. Injectable click-crosslinked hydrogel containing resveratrol to improve the therapeutic effect in triple negative breast cancer. *Mater. Today Bio* **2022**, *16*, 100386. [CrossRef]
130. Bruna, L.; Melo, R.; Cátia, G.A.; André, F.; Moreira, I.J.; Correia, I.J.; de Melo-Diogo, D. Chitosan-based injectable in situ forming hydrogels containing dopamine-reduced graphene oxide and resveratrol for breast cancer chemo-photothermal therapy. *Biochem. Eng. J.* **2022**, *185*, 108529.
131. Kotta, S.; Aldawsari, H.M.; Badr-Eldin, S.M.; Nair, A.B.; Kaleem, M.; Dalhat, M.H. Thermosensitive Hydrogels Loaded with Resveratrol Nanoemulsion: Formulation Optimization by Central Composite Design and Evaluation in MCF-7 Human Breast Cancer Cell Lines. *Gels* **2022**, *8*, 450. [CrossRef]
132. Hung, C.F.; Lin, Y.K.; Huang, Z.R.; Fang, J.Y. Delivery of resveratrol, a red wine polyphenol, from solutions and hydrogels via the skin. *Biol. Pharm. Bull.* **2008**, *31*, 955–962. [CrossRef]

133. Paulo, L.; Ferreira, S.; Gallardo, E.; Queiroz, J.A.; Domingues, F. Antimicrobial activity and effects of resveratrol on human pathogenic bacteria. *World J. Microbiol. Biotechnol.* **2010**, *26*, 1533–1538. [CrossRef]
134. Zykova, T.A.; Zhu, F.; Zhai, X.; Ma, W.Y.; Ermakova, S.P.; Lee, K.W.; Bode, A.M.; Dong, Z. Resveratrol directly targets COX-2 to inhibit carcinogenesis. *Mol. Carcinog.* **2008**, *47*, 797–805. [CrossRef] [PubMed]
135. Varoni, E.M.; Lo Faro, A.F.; Sharifi-Rad, J.; Iriti, M. Anticancer molecular mechanisms of resveratrol. *Front. Nutr.* **2016**, *3*, 8. [CrossRef] [PubMed]
136. Pezzuto, J.M. Resveratrol as an inhibitor of carcinogenesis. *Pharm. Biol.* **2008**, *46*, 443–573. [CrossRef]
137. Chen, J.; Liu, H.; Zhao, C.; Qin, G.; Xi, G.; Li, T.; Wang, X.; Chen, T. One-step reduction and PEGylation of graphene oxide for photothermally controlled drug delivery. *Biomaterials* **2014**, *35*, 4986–4995. [CrossRef]
138. Yang, T.; Ren, H.; Zhang, W.; Rong, L.; Zhang, D. Resveratrol-Coated Gold Nanoflowers for CT Imaging and Apoptosis/Photothermal Synergistic Therapy of Malignant Melanoma. *ACS Omega* **2023**, *8*, 34629–34639. [CrossRef]
139. Xiang, S.; Zhang, K.; Yang, G.; Gao, D.; Zeng, C.; He, M. Mitochondria-Targeted and Resveratrol-Loaded Dual-Function Titanium Disulfide Nanosheets for Photothermal-Triggered Tumor Chemotherapy. *Nanoscale Res. Lett.* **2019**, *14*, 211. [CrossRef]
140. Shaito, A.; Posadino, A.M.; Younes, N.; Hasan, H.; Halabi, S.; Alhababi, D.; Al-Mohannadi, A.; Abdel-Rahman, W.M.; Eid, A.H.; Nasrallah, G.K.; et al. Potential Adverse Effects of Resveratrol: A Literature Review. *Int. J. Mol. Sci.* **2020**, *21*, 2084. [CrossRef]
141. Li, X.; Chen, L.; Huang, M.; Zeng, S.; Zheng, J.; Peng, S.; Wang, Y.; Cheng, H.; Li, S. Innovative strategies for photodynamic therapy against hypoxic tumor. *Asian J. Pharm. Sci.* **2023**, *18*, 100775. [CrossRef]
142. Olszowy, M.; Nowak-Perlak, M.; Woźniak, M. Current Strategies in Photodynamic Therapy (PDT) and Photodynamic Diagnostics (PDD) and the Future Potential of Nanotechnology in Cancer Treatment. *Pharmaceutics* **2023**, *15*, 1712. [CrossRef]
143. Li, R.T.; Zhu, Y.D.; Li, W.Y.; Hou, Y.K.; Zou, Y.M.; Zhao, Y.H.; Zou, Q.; Zhang, W.H.; Chen, J.X. Synergistic photothermal-photodynamic-chemotherapy toward breast cancer based on a liposome-coated core-shell AuNS@NMOFs nanocomposite encapsulated with gambogic acid. *J. Nanobiotechnology* **2022**, *20*, 212. [CrossRef]
144. Overchuk, M.; Weersink, R.A.; Wilson, B.C.; Zheng, G. Photodynamic and Photothermal Therapies: Synergy Opportunities for Nanomedicine. *ACS Nano* **2023**, *17*, 7979–8003. [CrossRef]

biomedicines

MDPI

Article

Cytocidal Effects of Interstitial Photodynamic Therapy Using Talaporfin Sodium and a Semiconductor Laser in a Rat Intracerebral Glioma Model

Yuki Saito [1], Shinjiro Fukami [1,*], Kenta Nagai [1], Emiyu Ogawa [2], Masahiko Kuroda [3], Michihiro Kohno [1] and Jiro Akimoto [1,4]

[1] Department of Neurosurgery, Tokyo Medical University, Tokyo 160-0023, Japan; yuki2712@tokyo-med.ac.jp (Y.S.); k-nagai@tokyo-med.ac.jp (K.N.); mkouno@tokyo-med.ac.jp (M.K.); jiroaki@gmail.com (J.A.)

[2] Department of Electronics and Electrical Engineering, Faculty of Science and Technology, Keio University, Yokohama 223-8522, Japan; emiyu@elec.keio.ac.jp

[3] Department of Molecular Pathology, Tokyo Medical University, Tokyo 160-8402, Japan; kuroda@tokyo-med.ac.jp

[4] Department of Neurosurgery, Kohsei Chuo General Hospital, Tokyo 153-8581, Japan

* Correspondence: fukami@tokyo-med.ac.jp; Tel.: +81-3-3342-6111 (ext. 5773); Fax: +81-3340-4285

Abstract: This preclinical study was conducted to investigate the efficacy of interstitial PDT (i-PDT) for malignant gliomas arising deep within the brain, which are difficult to remove. C6 glioma cells were implanted into the basal ganglia of rats, and 3 weeks later, the second-generation photosensitizer talaporfin sodium (TPS) was administered intraperitoneally. Ninety minutes after administration, a prototype fine plastic optical fiber was punctured into the tumor tissue, and semiconductor laser light was irradiated into the tumor from a 2-mm cylindrical light-emitting source under various conditions. The brain was removed 24 h after the i-PDT and analyzed pathologically. The optical fiber was able to puncture the tumor center in all cases, enabling i-PDT to be performed. Histological analysis showed that tumor necrosis was induced in areas close to the light source, correlating with the irradiation energy dose, whereas apoptosis was induced at some distance from the light source. Irradiation using high energy levels resulted in tissue swelling from strong tumor necrosis, and irradiation at 75 J/cm^2 was most suitable for inducing apoptosis. An experimental system of i-PDT using TPS was established using malignant glioma cells transplanted into the rat brain. Tumor cell death, which correlated with the light propagation, was induced in tumor tissue.

Keywords: interstitial photodynamic therapy; talaporfin sodium; malignant glioma; optical fiber; apoptosis

1. Introduction

Glioblastoma, which is the most common type of malignant brain tumor, is a rare disease occurring in about 3 per 100,000 people [1,2]. The prognosis is highly unfavorable, with a median overall survival (m-OS) of 14.6 months and a 5-year survival rate of 9.8%, even with the standard treatment of maximal surgical resection, 60-Gy radiotherapy, and chemotherapy [3]. The rate of tumor removal by craniotomy is one of the most important prognostic factors, and for glioblastoma, a removal of more than 78% is considered to be the threshold for a comparatively favorable prognosis [4–8]. The prognosis for patients with residual tumors of 10 cm^3 or less after surgical removal is 32% at 6 months, whereas for tumors of 15 cm^3 or more, it is 3% at 6 months [9]; hence, the prognosis is highly unfavorable in patients in whom surgical removal is not feasible.

Photodynamic therapy (PDT) is a novel therapy being investigated for these malignant gliomas with a highly unfavorable prognosis. PDT, which has been the subject of clinical research since the 1980s, is a method of treatment in which a photosensitizer (PS), administered either orally or intravenously, is selectively taken up by tumors. By exposing the tumor to laser light of a specific wavelength, the PS taken up by the tumor cells is excited and the dissolved oxygen in the tissue is converted into highly toxic singlet oxygen, thereby producing an anti-tumor effect [10–17]. In Japan, an investigator-initiated clinical trial of PDT for malignant brain tumors using the second-generation PS, talaporfin sodium (TPS), started in 2009 and demonstrated an m-OS of 24.8 months and a median progression-free survival (m-PFS) of 12 months in patients with newly diagnosed glioblastoma, leading to its insurance approval in 2013 [17–19]. Subsequent reports by Nitta et al. also showed an m-OS of 27.4 months and an m-PFS of 19.6 months, surpassing previous treatment results, demonstrating the efficacy of PDT for newly diagnosed glioblastoma [20]. However, the current conventional method of PDT is to apply laser irradiation to the resection cavity wall after maximal safe resection [17–20], which makes it difficult to perform PDT in difficult-to-resect areas such as the basal ganglia and brainstem, and hence, does not contribute to improving the prognosis of these patients. Indeed, the m-OS for patients with high-grade basal ganglia gliomas is 10.0 months [21,22]; for those with diffuse intrinsic pontine gliomas, this is 9 to 16 months [22–25], with a 1-year survival rate of 34.9% [25]; and for glioblastomas, which are very severe, the m-OS is 6.7 months [26]. Increasing the removal rate of tumors in this region to improve prognoses as much as possible is extremely difficult in terms of preserving neurological function, with reports of adverse events resulting from surgery in as many as 34% of patients [27,28]. In other words, in cases of patients with malignant gliomas in the basal ganglia or brainstem, at present, surgery is performed only for a histological diagnosis and no treatment has yet been developed that can improve the prognosis.

Many researchers have investigated treatments using PDT to improve the prognosis of malignant gliomas arising in brain regions that are difficult to treat. The most promising of these is intra-tissue PDT (interstitial PDT: i-PDT), in which a thin fiber is punctured into the tumor and laser light is emitted from inside the tumor tissue. Although the optical delivery and irradiation protocol of PDT using fine fibers requires full investigation, the efficacy of i-PDT using a hematoporphyrin ester for malignant gliomas arising in the basal ganglia has already been reported by Kaneko in Japan [29]. Furthermore, in Germany, a prospective clinical study was reported on the efficacy and safety of i-PDT for recurrent malignant gliomas, in which 5-aminolevulinic acid was administered and an optical fiber called a laser diffuser was punctured into the tumor at several points [30–32]. Each of these reports demonstrates the potential of this new treatment for malignant gliomas arising in difficult-to-resect regions.

We previously investigated the potential of i-PDT with the photosensitizer TPS for malignant gliomas using a C6 glioma subcutaneous tumor model in nude mice [33]. We were able to successfully estimate the optimal timing of i-PDT from the administration of TPS, develop a fiber device for the i-PDT, and determine the optimal conditions for i-PDT to cause tumor tissue injury [33]. In the present study, we report the establishment of a model in which C6 glioma cells were transplanted into the brains of immunocompetent rats, to create a model that more closely mimics glioblastoma patients than the nude mouse subcutaneous tumor model, and to investigate whether i-PDT is as effective in this model as in the subcutaneous transplant model.

2. Material and Methods

2.1. Materials

2.1.1. Cell Culture

Culture medium was prepared by adding 50 mL of heat-inactivated 10% fetal bovine serum (Gibco Thermo-Fisher Scientific, Waltham, MA, USA) and 50 U/mL penicillin–streptomycin solution (Sigma-Aldrich Co., LLC., St. Louis, MO, USA) to 500 mL of Dulbecco's Modified Eagle Mediu (Gibco Thermo-Fisher Scientific). The C6 rat glioma cell line (Riken Cell Bank, Tsukuba, Ibaraki, Japan) was cultured in this culture medium in a

10-cm diameter dish in a CO_2 incubator (5% CO_2 concentration, 37 °C). The cells became confluent in about 3 days, and passages were repeated as necessary.

2.1.2. Rat Basal Ganglia C6 Transplantation Model

C6 glioma cells were cultured under the conditions described above and those that reached confluency were detached from the culture dish with 1 mL of trypsin (Sigma-Aldrich Co. LLC.) and collected. The resulting cell suspension was centrifuged at 27 °C, 1000 rpm, for 3 min. After removal of the supernatant, the volume was adjusted to $2 \times 10^4/\mu L$ with phosphate-buffered saline for transplantation into rats. Eight-week-old male Wistar rats (Japan SLC Inc., Hamamatsu, Shizuoka, Japan) weighing 200 g to 250 g were sedated under intraperitoneal anesthesia with a triad of anesthetics. Then, the head was fixed to a stereotactic instrument for rats (SR-5R-HT: Narishige Scientific Instruments Lab., Tokyo, Japan), a dental drill was used to puncture a point 3 mm right lateral and 2 mm anterior to Bregma, and a tumor cell suspension was injected into the striatum at a depth of 5 mm from the brain surface at this point. Ten microliters of cell suspension was slowly injected using a 27-gauge—gauge Hamilton micro-syringe, the cells were transplanted, and the mice were housed under normal conditions for 3 weeks. For the operation, a jig with a 27-gauge needle threading was fabricated to allow implantation at a certain depth from the brain surface. A special jig (created using a 3D printer) that can be implanted intracranially was used to reduce brain contusions that could occur during subsequent manipulations, while also making it easy to identify the tip of the emission site of the optical fiber on Hematoxylin and Eosin (HE) specimens (Figure 1A–C).

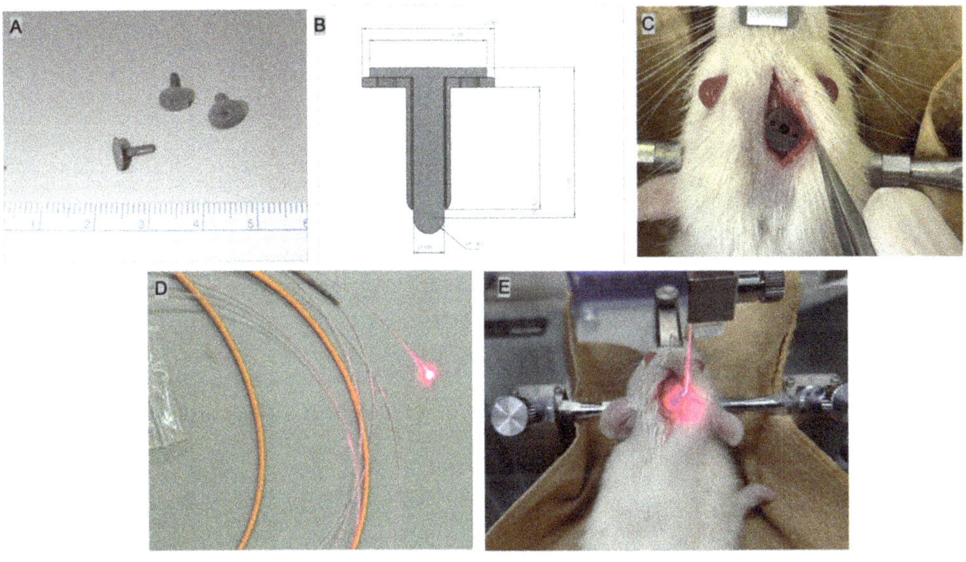

Figure 1. Devices and method for performing i-PDT. (**A**) Jig for accurately guiding the optical fibers, created by the Authors, using a 3D printer. (**B**) As for the size of the jig, the total length is 7.7 mm, the diameter of the fixed blade is 7 mm, and the diameter of the optical fiber passage hole is 1.6 mm. (**C**) Jig on the surface of rat brain. (**D**) A thin plastic optical fiber that emits cylindrical light with a long diameter of 2-mm, which was prepared by the Authors. (**E**) Practical application of i-PDT to rat brain tumors.

2.1.3. Talaporfin Sodium (TPS)

TPS, with a molecular weight of 799.69, is a second-generation water-soluble PS consisting of an aspartic acid amide bonded to a chlorin backbone derived from plant chlorophyll. It is characterized by faster elimination and a shorter period of post-administration shading than the first-generation PS. Excitation wavelengths are 405 nm and 664 nm, and the latter wavelength is used for PDT as this wavelength is not affected by absorption by red blood cells and water. In Japan, TPS was first covered by insurance for early-stage lung cancer in 2004, followed by malignant glioma in 2013, and recurrent esophageal cancer after radio-chemotherapy in 2015; thereby, it became widely used in Japan as a PS for PDT.

2.1.4. Laser Irradiation Fiber Probe

The fine-diameter optical fiber used for cylindrical irradiation is an original polymethyl methacrylate optic fiber (diameter: 0.8 mm; emission band: 2.0 mm; emission area: 0.0553 cm^2; transmittance: 65%; Nissei Electric Co., Machida, Tokyo, Japan) (Figure 1D). The laser light source Rouge-LD (664 nm, 500 mW, class 4 continuous-wave laser diode, Cyber Laser Inc., Tokyo, Japan) was used. To ensure stable and accurate implantation of a thin optical fiber at the tumor center, an original experimental jig was developed using a 14 G intravenous needle (Becton Dickinson and Company, Inc., Franklin Lakes, NJ, USA) with a 3D-printed fixing blade attached with instant adhesive (Figure 1A–C). A stopper was provided behind the optical fiber to ensure that the light-emitting surface of the thin optical fiber was exposed at the tip of the outer cylinder so that the fiber could always be irradiated accurately to the tumor center.

2.2. Methods

2.2.1. Implementation of i-PDT

Tumors grown in the right basal ganglia of disease-model rats (N = 5) were irradiated. Ten milligrams per kilogram dose of TPS was administered intraperitoneally to rats. After 90 min, an optical fiber was inserted 10 mm through the guidance hole of the dedicated jig, and red laser light at a wavelength of 664 nm was irradiated into the tumor tissue under various conditions (Figure 1E and Table 1). Groups A to C were the i-PDT groups (A: 150 J/cm^2; B: 75 J/cm^2; C: 50 J/cm^2), group D was treated with intraperitoneal TPS but no laser irradiation, and group E received neither TPS nor laser irradiation. The special jig and rat scalp were firmly fixed with tissue instant adhesive and 4-0 nylon thread to prevent them from falling off (Figure 1E). During i-PDT, induction anesthesia was performed using intraperitoneal pentobarbital and inhalation anesthesia with isoflurane (Pfizer Inc., New York, NY, USA).

Table 1. Conditions for performing i-PDT in each group.

Group	N	Energy Density (J/cm^2)	Power Density (mW/cm^2)	Input Power (mW)	Time (Second)	TPS
A	5	150	150	8.3	1025	(+)
B	5	75	73	4	1025	(+)
C	5	50	150	8.3	334	(+)
D	5	0	0	0	0	(+)
E	5	50	150	8.3	334	(−)

2.2.2. Post-Irradiation Tissue Sampling

Twenty-four hours after i-PDT, the treated rats were deeply anesthetized with a triad of anesthetics and sacrificed by cervical dislocation. The cranium was opened, and the brain was removed in a single block and then fixed in 10% formalin solution. After sufficient fixation, the fiber insertion corridor was cut out in the coronal plane so that it could be observed, and the tumor surface was exposed and paraffin-embedded.

2.2.3. Histopathological Analyses

Morphological assessments were performed by HE staining. To investigate the tumor-killing effect of i-PDT using TPS, the TUNEL staining method was used (in situ apoptosis detection kit MK500, Takara Bio, Tokyo, Japan). The label was a rabbit-derived anti-fluorescein isothiocyanate horseradish peroxidase conjugate with 10× dilution of terminal nucleotidyl transferase, and the chromogenic substrate was a 50× dilution of 3,3′-diaminobenzidine and the ApopMark™ Apoptosis Detection Kit (Exalpha Biologicals Inc., Maynard, MA, USA). Phospho-tungstic acid hematoxylin (PTAH) staining and immunostaining with an anti-CD31 antibody were used to assess thrombi in tumor vessels. For further immunological evaluation, immunohistochemical analyses were performed using monoclonal antibodies for CD4, CD8, CD20, GFAP, Iba-1, TP53, and CD68 (Abcam plc., Cambridge, UK). Analysis of each virtual slide was performed using NDP.viewer2 U12388-01 (Hamamatsu Photonics K.K., Hamamatsu, Shizuoka, Japan). For specimens stained with the ApopMark™ DNA Fragmentation Detection Kit, 10 locations around the tip of fiber used for laser irradiation of the tumor were arbitrarily selected in the 200× field-of-view of sections from each condition, and the number of cells positive for staining was counted using Patholoscope software (Mitani Co., Tokyo, Japan) to determine whether there are significant differences between groups.

2.2.4. Statistical Processing

All data were analyzed using IBM SPSS Statistics 29.0.2 software (Statistical Package for the Social Sciences; IBM Japan, Ltd., Tokyo, Japan). Statistical analysis was performed by one-way analysis of variance (ANOVA), and a p-value less than 0.05 was considered to indicate a statistically significant difference between groups.

3. Results

3.1. *C6 Transplantation Rat Basal Ganglia Glioma Model*

In all rats, a tumor with a mass of 3 to 5 mm was formed in the right basal ganglia, with tumor cell infiltration at the border with the normal brain. Necrosis and hemorrhagic changes were observed in the center of the tumor, and the cells comprising the tumor were markedly atypical, with abundant mitotic figures and neovascularization with the proliferation of endothelial cells. Large tumors were also formed with edema in the surrounding brain and a strong deviation of the median structure, closely mimicking the glioblastoma pathology observed in clinical practice.

3.2. *Tumor Pathology 24 h after i-PDT*

HE staining demonstrated tumor-destructive changes in each of the i-PDT irradiation conditions, although there were some differences. In group A (150 J/cm^2)—in which tumors were irradiated with the highest energy density—tumor tissues showed extensive necrotic changes, combined with edematous changes in the surrounding brain, demonstrating a deviation of the median structure. Extensive necrotic changes in tumor cells were also induced in groups B (75 J/cm^2) and C (50 J/cm^2), but residual tumor tissues were observed, with no edematous changes that could cause a deviation of the midline structure. The control groups D and E also showed necrotic changes in the tumor tissue, but only in some parts of the tissue, and did not show extensive tumor tissue necrosis as observed in the i-PDT group (A–C) (Figure 2). In the assessment of intra-tumor vessels, fibrin thrombus formation as identified by PTAH staining was not observed in the i-PDT-treated groups (A–C) but CD31 immunostaining gave the impression of a narrow vascular bed in the i-PDT-treated groups that was not observed in the control group (D, E) (Figure 3G,H).

Various immunostaining evaluations were performed, and staining with an anti-GFAP antibody revealed scattered tumor cells in the i-PDT group outside of the necrotic area around the light source, which had a spherical morphology with shortened tumor cell processes and a condensed cytoplasm (Figure 3A,B). Staining with a TP53 antibody demonstrated a high rate of positive cells in the same area (Figure 3C,D). Evaluation of apoptosis by TUNEL staining showed numerous positive cells outside the necrotic area around the light source in the i-PDT-treated groups but almost no positive cells in the control groups (Figure 3E,F). No significant differences were found between the i-PDT and control groups in terms of the positivity rate of immunocompetent cells, such as those positive for CD4, CD8, Iba-1, and CD68.

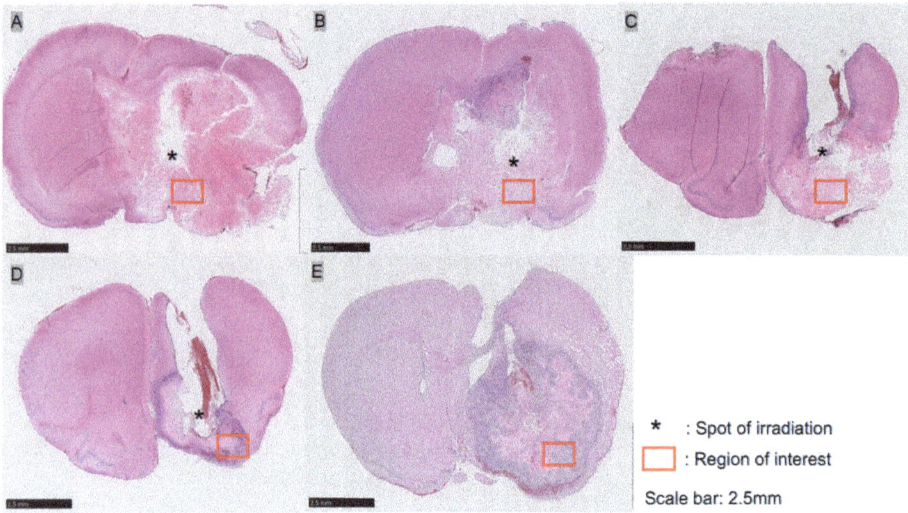

Figure 2. HE-stained macroscopic view of C6 glioma tissue after undergoing i-PDT of various conditions: (**A**) 150 J/cm^2; (**B**) 75 J/cm^2; (**C**) 50 J/cm^2; (**D**,**E**) control. In group A, tumor tissue swelling occurred along with extensive destruction of tissue owing to i-PDT, resulting in the deviation of the midline structures. In groups B and C, extensive destruction of tumor tissue occurred, except in the spots of light irradiation (asterisk). Group D of the control showed necrosis in the center of the tumor but with viable tumor tissue around it. Group E showed massive tumor tissue with no necrosis or apoptosis. The red boxed area in each figure is the region of interest where detailed pathological analysis was subsequently performed. The scale bar for all figures indicates a length of 2.5 mm.

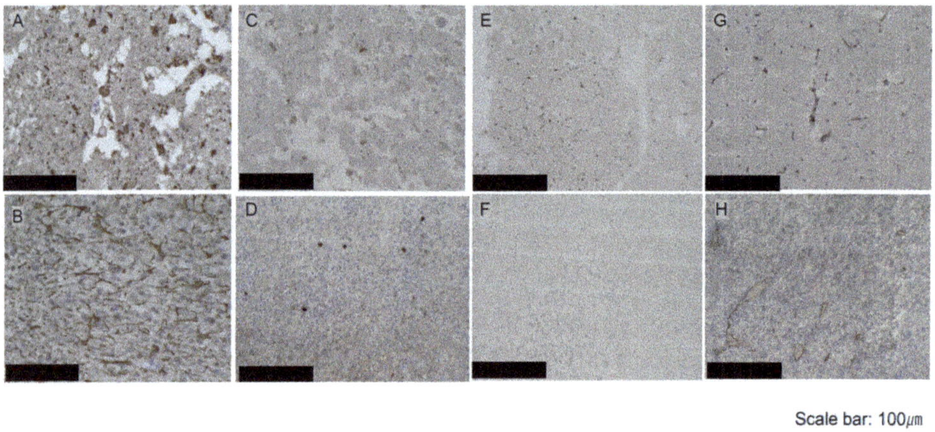

Figure 3. Microscopic view of C6 glioma tissue slightly distant from the light source (ROI in Figure 2) in the i-PDT (75 J/cm^2) and control groups. (**A**,**B**) Immunostaining of the tumor tissue with an anti-GFAP antibody (100×). i-PDT resulted in shortening of tumor cell processes and condensation of their cytoplasm. (**C**,**D**) Immunostaining with an anti-p53 antibody (100×). Higher p53-positive cell rates were observed in the i-PDT-treated groups than in the control groups. (**E**,**F**) Assessment of apoptosis by TUNEL staining (100×). More TUNEL-positive apoptotic cells were observed in the i-PDT-treated groups than in the control groups. (**G**,**H**) Immunostaining with an anti-CD31 antibody (100×). Narrowing of the vascular bed of CD31-positive tumor vessels was observed in the i-PDT-treated groups. The scale bar for all figures indicates a length of 100 μm.

3.3. Quantitative Assessment of Apoptosis with ApopMark™

Specimens stained with ApopMark™ from the i-PDT-treated groups (A: 150 J/cm^2: B: 75 J/cm^2, and C: 50 J/cm^2) and the control groups were captured as virtual slides, and 10 regions of interest were randomly selected from around the optical diffuser of the laser-irradiated fiber. The number of apoptosis-induced cells in the 200× field-of-view was measured using Pathologscope software. The results showed that the i-PDT-treated groups had significantly more apoptosis-induced cells than the control groups, with the 75 J/cm^2 irradiation group having the most apoptosis-induced cells among i-PDT-treated groups (A–C), although no statistically significant differences were observed (Figures 4 and 5).

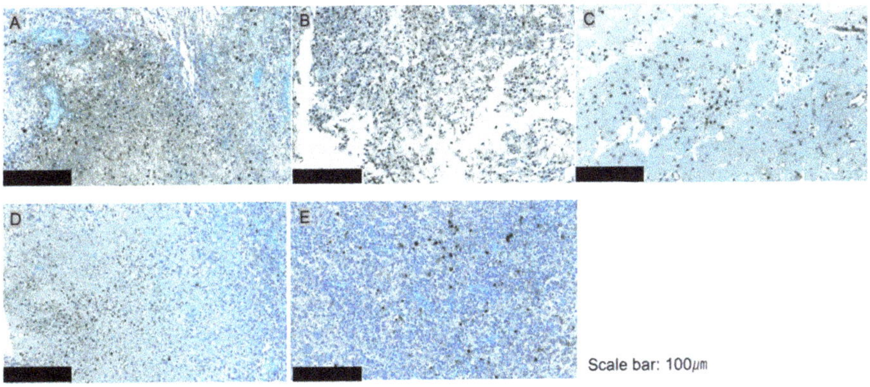

Figure 4. Apoptosis assessment of i-PDT-treated (**A–C**) and control (**D**,**E**) groups in ROIs of Figure 2 by ApopMark™ (100×). More ApopMark™-positive apoptotic cells are seen in the i-PDT groups than in the control groups, with the highest positive cell rate in the 75 J/cm^2 treatment group (**B**). The scale bar for all figures indicates a length of 100 μm.

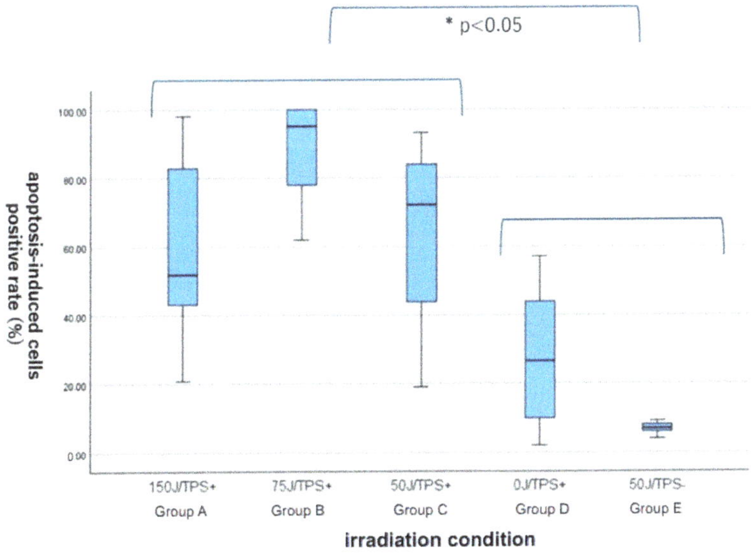

Figure 5. Comparison of the number of ApopMark™-positive cells in the i-PDT-treated groups (**A–C**) and the control groups (**D**,**E**). There were statistically significantly more apoptotic cells in the i-PDT-treated groups than in the control groups but there were no statistically significant differences among the 3 i-PDT-treated groups.

4. Discussion

PDT, a treatment for malignant brain tumors, is considered to be an additional treatment that targets residual infiltrating cells after maximal safe resection by craniotomy, and there have been many reports on its clinical efficacy and safety [17–20,34–36]. In other words, at present, PDT is only indicated for patients in whom near-total removal of the tumor is judged to be possible before surgery. However, a review of the location of 2049 glioblastomas reported in the Japanese Brain Tumor Registry (2005–2008) showed that 61 glioblastomas were located in the basal ganglia, 85 in the thalamus and hypothalamus, and 38 in the brainstem [37]. This means that 184 (8.98%) glioblastomas occurred in these difficult-to-resect sites, and these cases were contra-indicated for PDT in the preoperative diagnosis. However, we have been investigating whether these patients with glioblastoma that occur in these site—which constitute less than 10% overall—could also benefit from PDT. We hence investigated the possibility of i-PDT as an alternative to conventional PDT, which irradiates the surface of the tumor resection cavity.

We previously demonstrated that PDT using TPS has tumor-cell-killing effects on several malignant glioma cell lines in an in vitro study, in a TPS concentration- and time-dependent manner after PDT [38–42]. It has also been reported that the cell-killing effect is essentially the result of both tumor cell necrosis and apoptosis caused by PDT-induced singlet oxygen toxicity, with lysosome and mitochondrial crosstalk in the tumor cytoplasm playing an important role [39,40,42]. Furthermore, in vivo experiments of surface-irradiated PDT in a rat brain tumor model in which C6 cells were implanted on the surface of the brain have shown that surface irradiation induces necrosis of tumor tissue on the surface of the brain close to the light-irradiated surface, and apoptosis deep within the brain, with the effect being enhanced over time [43]. In other words, it was speculated that the induction of a cell-killing effect similar to surface irradiation could be achieved for deep-brain tumors if light irradiation could be applied precisely after TPS administration.

The problem was how to deliver the light into the tumor tissue. We therefore used a light diffuser made of plastic optical fibers with a diameter of 250 μm, which had already been developed by co-authors Ogawa et al. [44]. This fiber has a 0.8-to-1-mm diameter fluorinated resin material was used as the outer structure, which is useful from the viewpoint of biocompatibility and heat resistance, through which a thin plastic optical fiber is threaded through to ensure rigidity to puncture the tumor. The light diffuser has a diffusion length of 2 mm to enable full circumferential irradiation, and is constructed to create an air layer between the fibers and the outer tube to reduce heat generation owing to prolonged irradiation [44]. In our previous study, we established an experimental system of i-PDT in a nude mouse C6 subcutaneous implant glioma model using this optical fiber and confirmed that accurate intra-tumor puncture and irradiation were possible and that tumor tissue destruction was achieved, correlating with the irradiation light energy intensity [33]. Therefore, it was concluded that it would be possible to perform PDT experiments with full circumferential irradiation of the tumor center using this optical fiber in the present intracerebral transplantation model.

We found that the fiber could be punctured precisely into the center of the tumor in the basal ganglia C6 transplantation model, and i-PDT could be performed under various light irradiation conditions. Furthermore, in the area close to the i-PDT light source—which provides full circumferential irradiation—extensive tissue necrosis was induced, whereas at some distance, apoptosis was induced, confirmed by the TUNEL staining and ApopMark™. In the aforementioned experiment on a brain surface glioma model using C6 glioma cells, the results of the laser irradiation of the tumor surface using TPS showed that of the C6 glioma tissue extending 2 mm into the brain parenchyma, extensive necrosis occurred from the surface to 1.5 mm, with 0.5 mm at the deep end remaining [43]. The apical tumor tissue had undergone apoptosis, as confirmed by M30-CytoDeath immunostaining [43]. In the present i-PDT experiments, necrosis was observed in the area close to the light source, and apoptosis was observed at some distance from the light source, suggesting there is a gradient in the effect of PDT owing to the light delivery into the brain tissue in i-PDT using

TPS. In future experimental systems of i-PDT, it will be necessary to verify the consistency between light delivery and the actual degree of tissue injury.

The i-PDT group was divided into three energy density groups, i.e., 150 J/cm^2 (group A), 75 J/cm^2 (group B), and 50 J/cm^2 (group C), and the pattern of tumor tissue injury was compared. In group A, extensive necrosis of the tumor tissue was observed, some cells undergoing apoptosis were found, and the midline structure was deviated owing to swelling of the tissue caused by the severe necrosis. Group B also showed necrosis around the light source, and the density of cells undergoing apoptosis in the surrounding area was the highest. Group C also showed necrosis and apoptosis but the frequency of apoptosis was not as high as in group B. To develop i-PDT for deep-seated malignant gliomas for clinical use in the future, it is necessary to consider the safety of i-PDT for the surrounding brain structures, to establish a protocol for i-PDT that does not cause severe necrosis induction owing to tumor swelling, and substantial tissue cell injury by apoptosis. In our previous experiments, in a C6 subcutaneous transplantation glioma model in nude mice, we found that when the same energy density was applied, irradiation at a low power density over a long time was more effective than a high power density over a short time in inducing apoptosis [33], and a similar tendency is thought to exist in the intracerebral transplantation model.

In our previous experiments on i-PDT, tumor tissue damage was also observed as indicated by the gradient of light propagation but we also noted that the effect of PDT on tumor vessels contributes to the tissue damage effect [33]. In other words, our results showed that thrombus formation in tumor vessels by PDT could induce secondary ischemic tissue damage [45,46]. However, in the present study using an intracerebral transplantation glioma model, no noticeable thrombus formation was observed in the vessels in the tumor tissue according to the differences in the condition settings of PDT, such as its power density and energy density. However, a study using vascular endothelial staining with CD31 antibodies showed a trend toward a narrowing of the vascular bed of intra-tumor vessels in the i-PDT-treated group. The effect of PDT on tumor vessels has generally been described as a vascular shutdown effect but there have been scattered reports of vasoconstriction and vasospasm being induced, which do not lead to shutdown, depending on the irradiation conditions [46]. Our i-PDT conditions were highly likely effective in inducing necrosis and apoptosis of tumor tissue but appeared to have minor effect on the tumor vessels. Few reports to date have discussed the effects of i-PDT on blood vessels, and this may be an area of focus for future research [45,46].

In recent years, damage-associated molecular patterns that develop in tumor cells as a result of PDT-induced tissue damage have been shown to stimulate host anti-tumor immunocompetence [47,48]. In our present experimental system, we repeatedly performed analyses by sacrificing the rats at 24 h after i-PDT but we were unable to show a clear trend in the activation of all immune systems, including T cells, B cells, and microglia. This suggests that it may be crucial to evaluate the activation of tumor immunity by i-PDT at a later timepoint than 24 h after its implementation.

5. Conclusions

In our experimental study of i-PDT using TPS for a C6 brain transplantation rat glioma model, our prototype thin plastic optical fiber was able to accurately guide the light source to the tumor center and deliver cylindrical laser irradiation with a diffusion length of 2 mm. Although the pattern of tissue injury varied according to the conditions under which i-PDT was performed, tumor tissue necrosis was induced in the area close to the light source, whereas apoptosis was induced at a slightly more distant site, suggesting that the light propagation gradient was consistent with tissue injury. Other effects of PDT, such as vascular shutdown effects and secondary activation of host anti-tumor immunity, may require different i-PDT protocols and future studies analyzing changes over long periods of time after i-PDT using TPS.

Author Contributions: Conceptualization, Y.S., S.F., K.N. and J.A.; Data curation, Y.S., K.N. and J.A.; Investigation, Y.S., K.N., M.K. (Masahiko Kuroda) and J.A.; Methodology, Y.S., E.O., M.K. (Masahiko Kuroda) and J.A.; Resources, E.O.; Validation, S.F. and J.A.; Writing—original draft, Y.S. and J.A.; Writing—review and editing, S.F., M.K. (Masahiko Kuroda) and M.K. (Michihiro Kohno). All authors have read and agreed to the published version of the manuscript.

Funding: This work was supported by JSPS KAKENHI Grant Number JP22K09218.

Institutional Review Board Statement: All procedures described in this article were conducted in compliance with the Declaration of Helsinki on animal experiments, as amended in 2002. The presented research was approved by the Ethics Committee for Animal Experiments of Tokyo Medical University (project approval no.: R3-0096).

Informed Consent Statement: Not applicable.

Data Availability Statement: The datasets used and/or analyzed during the current study available from the corresponding author on reasonable request.

Acknowledgments: The authors are indebted to Helena Akiko Popiel, Center for International Education and Research of Tokyo Medical University, for her editorial review of the manuscript.

Conflicts of Interest: The authors declare that they have no conflicts of interest and no personal or institutional financial interests regarding any of the drugs, materials, or devices described in this article.

References

1. Ostrom, Q.T.; Bauchet, L.; Davis, F.G.; Deltour, I.; Fisher, J.L.; Langer, C.E.; Pekmezci, M.; Schwartzbaum, J.A.; Turner, M.C.; Walsh, K.M.; et al. The epidemiology of glioma in sdults: A "state of the science" review. *Neuro-Oncol.* **2014**, *16*, 896–913. [CrossRef] [PubMed]
2. Ostrom, Q.T.; Gittleman, H.; Truitt, G.; Boscia, A.; Krunko, C.; Barnholtz-Sloan, J.S. CBTRUS statistical report: Primary brain and central nervous system tumors diagnosed in united states in 2011–2015. *Neuro-Oncol.* **2018**, *20*, iv1–iv86. [CrossRef] [PubMed]
3. Stupp, R.; Mason, W.P.; van den Bent, M.J.; Weller, M.; Fisher, B.; Taphoorn, M.J.B.; Belanger, K.; Brandes, A.A.; Marosi, C.; Bogdahn, U.; et al. Radiotherapy plus concomitant and adjuvant temozolomide for glioblastoma. *N. Engl. J. Med.* **2005**, *10*, 987–996. [CrossRef] [PubMed]
4. Lacroix, M.; Abi-Said, D.; Fourney, D.R.; Gokaslan, W.; DeMonte, F.; Lang, F.F.; McCutcheon, E.; Hassenbusch, S.J.; Hokkand, E.; Hess, K.; et al. A multivariate analysis of 416 patients with glioblastoma multiforme: Prognosis, extent of resection, and survival. *J. Neurosurg.* **2001**, *95*, 190–198. [CrossRef] [PubMed]
5. Sanai, N.; Polly, M.Y.; McDermott, M.W.; Parsa, A.T.; Berger, M. An extent of resection threshold for newly diagnosed glioblastomas. *J. Neurosurg.* **2011**, *115*, 3–8. [CrossRef]
6. Sales, A.H.; Beck, J.; Schnell, O.; Fung, C.; Meyer, B.; Gempt, J. Surgical treatment of glioblastoma: State-of-the-art and future trend. *J. Clin. Med.* **2022**, *11*, 5354. [CrossRef] [PubMed]
7. Jusue-Torres, I.; Lee, J.; Germanwala, A.V.; Burns, T.; Parney, I.F. Effect of extent of resection on survival of patients with glioblastoma, IDH Wild-type, WHO Grade 4 (WHO 2021): Systematic review and meta-analysis. *World Neurosurg.* **2023**, *171*, e524–e532. [CrossRef] [PubMed]
8. Karshnia, P.; Dono, A.; Young, J.S.; Juenger, S.T.; Teske, N.; Hani, L.; Sciortino, T.; Mau, C.Y.; Bruno, F.; Nunez, L.; et al. Prognostic validation of a new classification system for extent of resection in glioblastoma: A report of RANO resect group. *Neuro-Oncol.* **2024**, *26*, 584–586. [CrossRef]
9. Keles, G.E.; Lamborn, K.R.; Chang, S.M.; Prados, M.D.; Berger, M.S. Volume of residual disease as a predictor of outcome in adult patients with recurrent supratentorial glioblastomas multiforme who are undergoing chemotherapy. *J. Neurosurg.* **2004**, *100*, 41–46. [CrossRef]
10. Castano, A.P.; Demidova, T.N.; Hambrin, M.R. Mechanism of photodynamic therapy: Part three- photosensitizer pharmacokinetics, biodistribution, tumor localization and modes of tumor destruction. *Photodiagn. Photodyn. Ther.* **2005**, *2*, 91–106. [CrossRef]
11. Quirk, B.J.; Brandal, G.; Donlon SVera, J.C.; Mang, T.S.; Foy ABLew, S.M.; Girotti, A.W.; Jogal, S.; LaViolette, P.S.; Connely, J.M.; Whelan, H.T. Photodynamic therapy (PDT) for malignant brain tumors-where do we stand? *Photodiagn. Photodyn. Ther.* **2015**, *12*, 530–544. [CrossRef] [PubMed]
12. Stylli, S.S.; Kaye, A.H.; MacGregor, L.; Howes, M.; Rajendra, P. Photodynamic therapy of high grade glioma-long term survival. *J. Clin. Neurosci.* **2005**, *12*, 389–398. [CrossRef]
13. Kostron, H.; Fiegele, T.; Akatuna, E. Combination of FOSCAN mediated fluorescence guided resection and photodynamic therapy as a therapeutic concept of malignant brain tumors. *Med. Laser Appl.* **2006**, *21*, 285–290. [CrossRef]
14. Stepp, H.; Beck, T.; Pongratz, T.; Meinel, T.; Kreth, F.W.; Tonn JCh Stummer, W. ALA and malignant glioma: Fluorescence-guided resection and photodynamic treatment. *J. Environ. Pathol. Toxicol. Oncol.* **2007**, *26*, 157–164. [CrossRef] [PubMed]

15. Muller, J.P.; Wilson, B.C. Photodynamic therapy of brain tumor—A work in progress. *Laser Surg. Med.* **2008**, *38*, 384–389. [CrossRef]
16. Eljamel, S.; Goodman, C.; Moseley, H. ALA and porphyrin fluorescence guided resection and repetitive PDT in glioblastoma multiforme: A single centre phase III randomized controlled trial. *Lasers Med. Sci.* **2008**, *23*, 361–367. [CrossRef]
17. Akimoto, J. Photodynamic therapy for malignant brain tumors. *Neurol. Med. Chir.* **2016**, *56*, 151–157. [CrossRef]
18. Akimoto, J.; Haraoka, J.; Aizawa, K. Preliminary clinical report on safety and efficacy of photodynamic therapy using talaporfin sodium for malignant gliomas. *Photodiagn. Photodyn. Ther.* **2012**, *9*, 91–99. [CrossRef]
19. Muragaki, Y.; Akimoto, J.; Maruyama, T.; Iseki, H.; Ikuta, S.; Nitta, M.; Maebayashi, K.; Saito, T.; Okada, Y.; Kaneko, S.; et al. Phase II clinical study on intraoperative photodynamic therapy with talaporfin sodium and semiconductor laser in patients with malignant brain tumors. *J. Neurosurg.* **2013**, *119*, 845–852. [CrossRef]
20. Nitta, M.; Muragaki, Y.; Maruyama, T.; Iseki, H.; Komori, T.; Ikuta, S.; Saito, T.; Yasuda, T.; Hosono, J.; Okamoto, S.; et al. Role of photodynamic therapy using talaporfin sodium and a semiconductor laser in patients with newly diagnosed glioblastoma. *J. Neurosurg.* **2018**, *131*, 1361–1368. [CrossRef]
21. Kim, Y.; Kudo, T.; Tamura, K.; Sumita, K.; Kobayashi, D.; Tanaka, Y.; Inaji, M.; Nariai, T.; Ishii, K.; Maehara, T. Clinical findings of thalamic and brain stem glioma including diffuse midline glioma, H3K27M mutant: A clinical study. *No Shinkei Geka* **2021**, *49*, 901–908. (In Japanese with English Abstract) [PubMed]
22. Dalmage, M.; LoPresti, M.A.; Sarkar, P.; Ranganathan, S.; Abdelmageed, S.; Pagadala, M.; Shlobin, N.A.; Lam, S.; DeCuypere, M. Survival and neurological outcomes after stereotactic biopsy of diffuse intrinsic pontine glioma: A systematic review. *J. Neurosurg. Pediatr.* **2023**, *32*, 665–672. [CrossRef] [PubMed]
23. Reyes-Botero, G.; Girty, M.; Mokhtari, K.; Labussiere, M.; Idbaith, A.; Delattre, J.Y.; Laigle-Donadey, F.; Sanson, M. Molecular analysis of diffuse intrinsic gliomas in adults. *J. Neurooncol.* **2014**, *116*, 405–411. [CrossRef] [PubMed]
24. Daoud, E.V.; Rajaram, V.; Cai, C.; Oberle, R.J.; Martin, G.R.; Raisanen, J.M.; White, C.L., 3rd; Foong, C.; Mickey, B.E.; Pan, E.; et al. Adult brain stem gliomas with H3K27M mutation: Radiology, pathology and prognosis. *J. Neuropathol. Exp. Neurol.* **2018**, *77*, 302–311. [CrossRef]
25. Dellaretti, M.; Reyns, N.; Touzet, G.; Dubois, F.; Gusmao, S.; Pereira, J.L.; Blond, S. Diffuse brain stem glioma: Prognostic factors. *J. Neurosurg.* **2012**, *117*, 810–814. [CrossRef]
26. Leibetseder, A.; Leitner, J.; Mair, M.J.; Meckel, S.; Hainfellner, J.A.; Aichholzer, M.; Widhalm, G.; Dickmann, K.; Weis, S.; Furtner, J.; et al. Prognostic factors in adult brainstem glioma: A tertiary care center analysis and review of the literature. *J. Neurol.* **2022**, *269*, 1574–1590. [CrossRef]
27. Faulkner, H.; Arnaout, O.; Hoshide, R.; Young, I.M.; Yeung, J.T.; Sughrue, M.E.; Teo, C. The surgical resection of brainstem glioma: Outcome and prognostic factors. *World Neurosurg.* **2021**, *146*, e639–e690. [CrossRef]
28. Barsouk, A.; Baldassari, M.P.; Khanna, O.; Andrews, C.E.; Ye, D.Y.; Velagapudi, L.; Al Saiegh, F.; Hafazalla, K.; Cunningham, E.; Patel, H.; et al. Glioblastoma with deep supratentorial extension is associated with a worse overall survival. *J. Clin. Neurosci.* **2021**, *93*, 82–87. [CrossRef]
29. Kaneko, S. Photodynamic therapy for human malignant glioma. *JJSLSM* **2011**, *32*, 131–138. (In Japanese with English Abstract) [CrossRef]
30. Beck, T.J.; Kreth, F.W.; Beyer, W.; Mehrkens, J.H.; Obermeier, A.; Stepp, H.; Stummer, W.; Baumgartner, R. Interstitial photodynamic therapy of nonresectable malignant glioma recurrence using 5-aminolevulinic acid induced protoporphyrin IX. *Laser Surg. Med.* **2007**, *39*, 386–393. [CrossRef]
31. Stepp, H.; Stummer, W. 5-ALA in the management of malignant glioma. *Laser Surg. Med.* **2018**, *50*, 399–419. [CrossRef] [PubMed]
32. Foglar, M.; Aumiller, M.; Bochmann, K.; Buchner, A.; El Fahim, M.; Quach, S.; Sroka, R.; Stepp, H.; Thon, N.; Forbrig, R. Interstitial photodynamic therapy of glioblastomas; A long-term follow-up analysis of survival and volumetric MRI data. *Cancers* **2023**, *15*, 2603. [CrossRef] [PubMed]
33. Nagai, K.; Akimoto, J.; Fukami, S.; Saito, Y.; Ogawa, E.; Takanashi, M.; Kuroda, M.; Kohno, M. Efficacy of interstitial photodynamic therapy using talaporfin sodium and a semiconductor laser for a mouse allograft glioma model. *Sci. Rep.* **2024**, *14*, 9137. [CrossRef]
34. Cramer, S.W.; Chen, C.C. Photodynamic therapy for treatment of glioblastoma. *Front. Surg.* **2020**, *6*, 81. [CrossRef] [PubMed]
35. Vermandel, M.; Dupont, C.; Lecomte, F.; Leroy, H.A.; Tuleasca, C.; Mordon, S.; Hadipanayis, C.G.; Reynes, N. Standardized intraoperative 5-ALA photodynamic therapy for newly diagnosed glioblastoma patients: A preliminary analysis of the INDYGO trial. *J. Neurooncol.* **2021**, *152*, 501–514. [CrossRef] [PubMed]
36. Mahmoudi, K.; Garvey, K.L.; Bouras, A.; Cramer, G.; Stepp, H.; Hesu Raj, J.G.; Bozec, D.; Busch, T.M.; Hadipanayis, C.G. 5-aminolevulinic acid photodynamic therapy for the treatment of high-grade gliomas. *J. Neurooncol.* **2019**, *141*, 595–607. [CrossRef]
37. Brain Tumor Registry of Japan (2005–2008). *Neurol. Med. Chir.* **2017**, *57*, 9–102. [CrossRef]
38. Tsutsumi, M.; Miki, Y.; Akimoto, J.; Haraoka, J.; Aizawa, K.; Hirano, K.; Beppu, M. Photodymamic therapy with talaporfin sodium induced dose-dependent apoptotic cell death in human glioma cell lines. *Photodiagn. Photodyn. Ther.* **2013**, *10*, 103–110. [CrossRef]
39. Miki, Y.; Akimoto, J.; Hiranuma, M.; Fujiwara, Y. Effect of talaporfin-mediated photodynamic therapy on cell death modalities in human glioblastoma T98G cells. *J. Toxicol. Sci.* **2014**, *39*, 821–827. [CrossRef]
40. Miki, Y.; Akimoto, J.; Yokoyama, S.; Homma, T.; Tsutsumi, M.; Haraoka, J.; Hirano, K.; Beppu, M. Photodynamic therapy in combination with talaporfin sodium induces mitochondrial apoptotic cell death accompanied with necrosis in glioma cells. *Biol. Pharm. Bull.* **2013**, *36*, 215–221. [CrossRef]

41. Miki, Y.; Akimoto, J.; Omata, H.; Moritake, K.; Hiranuma, M.; Hironaka, C.; Fujiwara, Y.; Beppu, M. Concomitant treatment with temozolomide enhances apoptotic cell death in glioma cells induced by photodynamic therapy with talaporfin sodium. *Photodiagn. Photodyn. Ther.* **2014**, *11*, 556–564. [CrossRef] [PubMed]
42. Miki, Y.; Akimoto, J.; Moritake, K.; Hironaka, C.; Fujiwara, Y. Photodynamic therapy using talaporfin sodium induces concentration-dependent programmed necroptosis in human glioblastoma T98G cells. *Lasers Med. Sci.* **2015**, *30*, 1739–1745. [CrossRef] [PubMed]
43. Namatame, H.; Akimoto, J.; Matsumura, H.; Haraoka, J.; Aizawa, K. Photodynamic therapy of C6-implanted glioma cells in the rat brain employing second-generation photosensitizer talaporfin sodium. *Photodiagn. Photodyn. Ther.* **2008**, *5*, 198–209. [CrossRef] [PubMed]
44. Ogawa, E.; Arai, T.; Usuda, J.; Otani, K.; Maehara, S.; Imai, K.; Kudo, Y.; Ono, S.; Ikeda, N. Photodynamic treatment using a small diameter diffused light irradiation probe made of plastic optical fiber. *JJSLM* **2020**, *41*, 25–29. (In Japanese with English Abstract) [CrossRef]
45. Moy, W.J.; Patel, S.J.; Lertsakdadet, B.S.; Arora, R.P.; Nielsen, K.M.; Kelly, K.M.; Choi, B. Preclinical in vivo evaluation of Npe6-mediated photodynamic therapy on normal vasculature. *Lasers Med. Sci.* **2012**, *44*, 158–162. [CrossRef]
46. McMahoon, K.S.; Wieman, J.; Moore, P.H.; Fingar, V.H. Effect of photodynamic therapy using mono-L-asparthyl chlorine e6 on vessel constriction, vessel leakage, and tumor response. *Cancer Res.* **1994**, *54*, 5374–5379.
47. Dudzik, T.; Domanski, I.; Makuch, S. The impact of photodynamic therapy on immune system in cancer-an update. *Front. Immunol.* **2024**, *15*, 1335920. [CrossRef]
48. Sasaki, M.; Tanaka, M.; Kojima, Y.; Nishie, H.; Shimura, T.; Kubota, E.; Kataoka, H. Anti-tumor immunity enhancement by photodynamic therapy with talaporfin sodium and anti-programmed death 1 antibody. *Mol. Ther. Oncolytics* **2023**, *28*, 118–131. [CrossRef]

biomedicines

Article

Effect of Clinicopathological Characteristics on the Outcomes of Topical 5-Aminolevulinic Acid Photodynamic Therapy in Patients with Cervical High-Grade Squamous Intraepithelial Lesions (HSIL/CIN2): A Retrospective Cohort Study

Yingting Wei [1,2,†], Jing Niu [1,2,†], Liying Gu [1,2], Zubei Hong [1,2], Zhouzhou Bao [1,2,3,*] and Lihua Qiu [1,2,3,*]

1 Department of Obstetrics and Gynecology, Ren Ji Hospital, School of Medicine, Shanghai Jiao Tong University, Shanghai 200127, China; tinawyting@126.com (Y.W.)
2 Shanghai Key Laboratory of Gynecologic Oncology, Ren Ji Hospital, School of Medicine, Shanghai Jiao Tong University, Shanghai 200127, China
3 State Key Laboratory of Oncogenes and Related Genes, Shanghai Cancer Institute, Ren Ji Hospital, School of Medicine, Shanghai Jiao Tong University, Shanghai 200127, China
* Correspondence: baozhouzhou@126.com (Z.B.); lilyqiulh@126.com (L.Q.); Tel.: +86-13482829929 (Z.B.); +86-21-68383809 (L.Q.)
† These authors contributed equally to this work.

Abstract: Background: Minimally-invasive 5-aminolevulinic acid photodynamic therapy (ALA-PDT) is used for treating cervical high-grade squamous intraepithelial lesions (HSIL/CIN2). The purpose of this study was to analyze the factors affecting the efficacy of ALA-PDT in the treatment of cervical HSIL/CIN2 in order to guide physicians in making appropriate treatment decisions. Methods: A retrospective study including 69 female patients with pathologically diagnosed HSIL/CIN2 was conducted. Patients were given six doses of 20% ALA-PDT at 7–14-day intervals. Cytology, HPV testing, colposcopy, and pathology were performed before treatment and at 6-month follow-up after treatment to assess efficacy. The main outcome of this study was the regression of HSIL/CIN2 and the clearance of high-risk HPV (hrHPV) infection after ALA-PDT treatment. Clinicopathological characteristics were collected to analyze the factors affecting the effectiveness of ALA-PDT treatment for HSIL/CIN2. Results: Between the successful and failed lesion regression group, there was a significant difference in sleeping disorders ($p < 0.05$). Between the successful and failed hrHPV clearance group, no statistically significant factors were found. With sensitivity values of 0.556 and 0.778, respectively, multivariate analysis showed that current smoking and sleeping disorders were independent prognostics of failure in lesion regression after ALA-PDT treatment. Conclusions: Smoking and sleep disorders were independent risk factors for failure in HSIL/CIN2 regression following ALA-PDT, suggesting the need for careful consideration of ALA-PDT for patients with these conditions.

Keywords: cervical high-grade squamous intraepithelial lesions; cervical intraepithelial neoplasia; 5-aminolevulinic acid; photodynamic therapy; high-risk HPV

1. Introduction

Cervical cancer ranks as the fourth most common malignancy among women, with approximately 600,000 new cases and 340,000 related deaths occurring globally each year [1]. The persistent infection with high-risk human papillomavirus (hrHPV) is strongly associated with the development of cervical cancer. While most hrHPV infections are transient and can be cleared by the immune system, persistent infections can lead to cervical intraepithelial neoplasia (CIN) and ultimately invasive cancer. CIN is classified based on the depth of cervical epithelial involvement into CIN1, CIN2, and CIN3, with CIN2 and CIN3 considered high-grade squamous intraepithelial lesions (HSIL). Each grade of CIN exhibits distinct biological behaviors, and the risk of progression varies. At least 12% of CIN3 cases will advance to invasive cancer, while 18% of CIN2 cases progress to CIN3 and invasive cancer [2,3]. Notably, 11% of untreated CIN2 patients under the age of 30 progress within two years [4]. Thus, timely intervention for patients with cervical HSIL is critical to prevent the onset of cervical cancer. The preferred treatment for cervical HSIL is excisional therapy, such as cold-knife conization or loop electrosurgical excision (LEEP). However, these procedures can lead to postoperative complications, including cervical stenosis, adhesions, cervical endometriosis, and cervical insufficiency, which may increase the risk of adverse pregnancy outcomes [5]. Therefore, for women with cervical HSIL who wish to preserve their fertility, a non-invasive therapeutic intervention that reduces the progression rate of CIN2 to CIN3+ is highly desirable.

In recent years, photodynamic therapy (PDT) has been gradually applied in the treatment of a variety of benign and malignant diseases [6,7]. PDT mediated by 5-aminolevulinic acid (5-ALA) is a minimally invasive treatment for CIN. Different from ablation or excision, ALA-PDT uses photodynamic responses to selectively destroy lesions for therapeutic purposes. Local application of ALA-PDT has the advantages of less trauma, less adverse reactions, good selectivity, repeatability, and rapid recovery. The histological remission rate of HSIL/CIN2 treated with ALA-PDT ranges from 77.8% to 91% [8]. Despite these promising results, some patients with HSIL do not benefit from ALA-PDT and may even progress to invasive cancer. Administering conservative treatment without thorough consideration can exacerbate their condition and miss the optimal treatment window.

Therefore, analyzing the factors influencing ALA-PDT outcomes in cervical HSIL patients is essential to identify the most effective treatment strategies. In this study, we collected data on 13 clinicopathological characteristics and investigated their potential predictive ability for outcomes in cervical HSIL patients following ALA-PDT.

2. Materials and Methods

2.1. Study Design and Study Population

This was a retrospective study. From March 2019 to October 2023, women with hrHPV infection, a pathologic diagnosis of cervical HSIL/CIN2, complete clinical data, and who had received ALA-PDT at the Gynecology Department of Renji Hospital, affiliated with Shanghai Jiao Tong University School of Medicine, were selected. All participants completed a questionnaire administered by a specialist physician. The questionnaire covered demographics such as age, smoking and drinking habits, HPV vaccination status, sleep disorders, number of births and miscarriages, as well as details regarding HPV infection type, cervical cytology results, pathology findings, and laboratory tests such as platelet–lymphocyte ratio (PLR) and neutrophil–lymphocyte ratio (NLR). Those with incomplete questionnaire information and follow-up results were excluded.

2.2. HPV Testing and Cytological Test

HPV genotyping was performed through the flow-through hybridization-based 21 HPV GenoArray Diagnostic Kit (Hybribio Ltd., Chaozhou, China) to identify and stratify HPV infection into the 15 high-risk types (HPV 16, 18, 31, 33, 35, 39, 45, 51, 52, 53, 56, 58, 59, 66, and 68) and six low-risk types (6, 11, 42, 43, 44, and 81). We classified HPV infections into the following two categories: HPV16/18 infections and non-HPV16/18 infections. Additionally, we differentiated between multiple-type HPV infections and single-type HPV infections. Cytological tests were performed using liquid-based cytology (ThinPrep cytological test, TCT). The results were categorized as negative for intraepithelial lesion or malignancy (NILM), atypical squamous cells of undetermined significance (ASCUS), low-grade squamous intraepithelial lesion (LSIL), HSIL, atypical squamous cells that could not exclude HSIL, and cervical cancer.

2.3. Photodynamic Therapy

To prepare the 20% ALA mixture, 354 mg of 5-ALA powder (Fudan Zhangjiang Bio-Pharm Co., Ltd., Shanghai, China) was combined with 1.5 mL of thermogel (stored at 4 °C). Sterile cotton pieces soaked in the ALA mixture were applied to the cervix and incubated for four hours in the dark. Subsequently, light irradiation at 635 nm and a dose of 100 J/cm^2 (LD600-C; Wuhan Yage Photo-Electronic Co., Ltd., Wuhan, China) was applied to the cervical surface and the endocervical canal for 30 min respectively. This treatment regimen was repeated six times at intervals of 7–14 days.

2.4. Follow-Up and Effect Evaluation

The primary outcomes measured were the regression of HSIL/CIN2 lesions and the clearance of hrHPV infection at the 6-month follow-up after ALA-PDT treatment. The specific grouping criteria were as follows:

1. Lesion: The Success Group was defined as regression of HSIL/CIN2, indicated by either a negative pathological diagnosis or a reduction in histological severity compared to the initial state after ALA-PDT treatment. The Failed Group was defined as persistence of HSIL/CIN2 (no change in histological diagnosis) or progression to CIN3+ at follow-up.
2. hrHPV: The Success Group was defined as conversion from baseline hrHPV-positive to hrHPV-negative at follow-up. The Failed Group was defined as hrHPV remaining positive at follow-up compared to the baseline.

2.5. Statistical Analysis

The collected data were analyzed using IBM SPSS software (Version 26.0, SPSS Inc., Chicago, IL, USA). Chi-square, Fisher's exact, and t tests were conducted to evaluate the difference in the incidence of different outcome groups, respectively. Univariate and multivariate analyses were utilized to find independent predictive factors related to the outcome of HSIL/CIN2 patients undergoing ALA-PDT.

2.6. Ethical Considerations

All procedures adhered to the ethical standards of the Institutional and National Research Committee as well as the 1964 Declaration of Helsinki and its subsequent amendments or comparable ethical standards. Informed consent was obtained from all participants included in the study. The study received approval from the Ethics Committee of Renji Hospital, affiliated with Shanghai Jiao Tong University School of Medicine (Committee's reference number KY2021-248-B).

3. Results

We enrolled 69 eligible patients with cervical HSIL/CIN2 who received ALA-PDT into this study and gathered 13 clinicopathological characteristics for analysis.

3.1. Analysis of Factors Influencing the Effect of ALA-PDT on HSIL/CIN2 Regression

Among the 69 patients, 60 had experienced histologic regression following ALA-PDT at the 6-month follow-up.

As shown in Table 1, the regression of HSIL/CIN2 was significantly more challenging in patients with sleep disorders compared to those without (65.0% vs. 95.9%, $p = 0.002$). Additionally, patients who were current smokers appeared to have a lower lesion regression rate compared to non-smokers (70.6% vs. 92.3%, $p = 0.058$). There was no significant difference between the two groups in terms of age, drinking history, HPV vaccination, history of childbirth or abortion, PLR, NLR, multiple HPV type infections, HPV16/18 infection, and TCT results ($p > 0.05$). Additionally, ALA-PDT combined with CO_2 lasers did not show a significantly different influence on the outcome compared to ALA-PDT alone ($p = 0.357$).

Table 1. Analysis of factors influencing the effect of ALA-PDT on HSIL/CIN2 regression.

Characteristics	Sum (n = 69)	Success in Lesion Regression (n = 60)	Failure in Lesion Regression (n = 9)	χ2/t	p/p'/ Fisher
Age [x ± s, years]	30.55 ± 7.25	30.63 ± 7.61	30.00 ± 4.44	0.243	0.809
NLR [x ± s]	2.14 ± 0.85	2.13 ± 0.89	2.20 ± 0.53	−0.207	0.837
PLR [x ± s]	128.75 ± 40.18	127.34 ± 40.37	138.18 ± 39.90	−0.752	0.455
current smoking [n, (%)]					
not	52 (75.4%)	48 (92.3%)	4 (7.7%)	3.586	0.058
yes	17 (24.6%)	12 (70.6%)	5 (29.4%)		
drinking history [n, (%)]					
not	58 (84.1%)	52 (89.7%)	6 (10.3%)	1.082	0.298
yes	11 (15.9%)	8 (72.7%)	3 (27.3%)		
sleeping disorders [n, (%)]					
not	49 (71%)	47 (95.9%)	2 (4.1%)	9.400	0.002
yes	20 (29%)	13 (65.0%)	7 (35.0%)		
HPV vaccination [n, (%)]					
not	54 (78.3%)	45 (83.3%)	9 (16.7%)	1.593	0.207
yes	15 (21.7%)	15 (100.0%)	0 (0.0%)		
HPV16/18 infection [n, (%)]					
not	25 (36.2%)	22 (88.0%)	3 (12.0%)	0.000	1.000
yes	44 (63.8%)	38 (86.4%)	6 (13.6%)		
multiple types of HPV infection [n, (%)]					
not	48 (69.6%)	43 (89.6%)	5 (10.4%)	0.349	0.554
yes	21 (30.4%)	17 (81.0%)	4 (19.0%)		
Thinpred cytologic test [n, (%)]					
≤ASCUS	40 (58%)	36 (90.0%)	4 (10.0%)	0.270	0.603
≥LSIL	29 (42%)	24 (82.8%)	5 (17.2%)		
history of giving birth [n, (%)]					
not	50 (72.5%)	43 (86.0%)	7 (14.0%)	0.000	1.000
yes	19 (27.5%)	17 (89.5%)	2 (10.5%)		
abortion history [n, (%)]					
not	48 (69.6%)	42 (87.5%)	6 (12.5%)	0.000	1.000
yes	21 (30.4%)	18 (85.7%)	3 (14.3%)		
combined with CO2 laser [n, (%)]					
not	44 (63.8%)	40 (90.9%)	4 (9.1%)	0.849	0.357
yes	25 (36.2%)	20 (80.0%)	5 (20.0%)		

NLR: neutrophil–lymphocyte ratio; PLR: platelets–lymphocyte ratio. Current smoking: a regular active smoker or exposure to environmental tobacco smoke. Non-current smokers had quit or never smoked regularly. Drinking history: participants engaged in drinking (≥14 drinks/wk for women) for a 30-day consecutive period and ≥2 heavy drinking days (defined as ≥4 drinks/d for women) in the 90 days preceding or during ALA-PDT. Sleep disorders: according to the International Classification of Sleep Disorders-3 (ICSD-3), insomnia disorders, sleep-related breathing disorders, central disorders of hypersomnolence, circadian rhythm sleep–wake disorders, sleep-related movement disorders, parasomnias, and other sleep disorders were defined as sleep disorders. HPV vaccination means a patient had received three shots of HPV vaccine before ALA-PDT, no matter whether the bivalent, quadrivalent, or nine-valent vaccine.

3.2. Analysis of Factors Influencing the Effect of ALA-PDT on hrHPV Clearance

Among the 69 patients, 44 achieved successful clearance of hrHPV, while 25 had persistent hrHPV infection at the 6-month follow-up. As shown in Table 2, current smoking was associated with a lower hrHPV clearance rate, but this difference was not statistically significant (p = 0.099). Additionally, results showed no statistically significant differences between the two groups for the other 12 clinicopathological characteristics.

Table 2. Analysis of factors influencing the effect of ALA-PDT on hrHPV clearance.

Characteristics	Sum (n = 69)	Success in hrHPV Clearance (n = 44)	Failure in hrHPV Clearance (n = 25)	χ^2/t	p/p′/ Fisher
Age [x ± s, years]	30.55 ± 7.25	30.77 ± 7.52	30.16 ± 6.89	0.335	0.739
NLR [x ± s]	2.14 ± 0.85	2.14 ± 0.84	2.15 ± 0.88	−0.083	0.934
PLR [x ± s]	128.75 ± 40.18	124.96 ± 40.93	135.42 ± 38.75	−1.040	0.302
current smoking [n, (%)]					
not	52 (75.4%)	36 (69.2%)	16 (30.8%)	2.726	0.099
yes	17 (24.6%)	8 (47.1%)	9 (52.9%)		
drinking history [n, (%)]					
not	58 (84.1%)	39 (67.2%)	19 (32.8%)	1.074	0.300
yes	11 (15.9%)	5 (45.5%)	6 (54.5%)		
sleeping disorders [n, (%)]					
not	49 (71.0%)	34 (69.4%)	15 (30.6%)	2.311	0.128
yes	20 (29.0%)	10 (50.0%)	10 (50.0%)		
HPV vaccination [n, (%)]					
not	54 (78.3%)	34 (63.0%)	20 (37.0%)	0.070	0.792
yes	15 (21.7%)	10 (66.7%)	5 (33.3%)		
HPV16/18 infection [n, (%)]					
not	25 (36.2%)	19 (76%)	6 (24%)	2.539	0.111
yes	44 (63.8%)	25 (58.6%)	19 (43.2%)		
multiple types of HPV infection [n, (%)]					
not	48 (69.6%)	32 (66.7%)	16 (33.3%)	0.574	0.449
yes	21 (30.4%)	12 (57.1%)	9 (42.9%)		
Thinpred cytologic test [n, (%)]					
≤ASCUS	40 (58.0%)	25 (62.5%)	15 (37.5%)	0.066	0.797
≥LSIL	29 (42.0%)	19 (65.5%)	10 (34.5%)		
history of giving birth [n, (%)]					
not	50 (72.5%)	33 (66.0%)	17 (34.0%)	0.391	0.532
yes	19 (27.5%)	11 (57.9%)	8 (42.1%)		
abortion history [n, (%)]					
not	48 (69.6%)	32 (66.7%)	16 (33.3%)	0.574	0.449
yes	21 (30.4%)	12 (57.1%)	9 (42.9%)		
combined with CO_2 laser [n, (%)]					
not	44 (63.8%)	28 (63.6%)	16 (36.4%)	0.001	0.976
yes	25 (36.2%)	16 (64.0%)	9 (36.0%)		

3.3. Univariate and Multivariate Regression Analyses of Factors Influencing the Efficacy of ALA-PDT

We further conducted univariate and multivariate regression analyses to identify independent prognostic factors for the efficacy of ALA-PDT in treating HSIL/CIN2. We employed a stepwise multiple linear regression model to analyze variables with a significance level of $p < 0.05$ from the univariate analysis. As shown in Table 3, first, univariate regression analysis was conducted, revealing associations between current smoking, sleep disorders, and lesion efficacy ($p < 0.05$). Subsequent multivariate regression analysis further indicated that both factors are independent risk factors for poorer histological prognosis ($p = 0.046$ and $p = 0.039$, respectively). Current smoking was strongly associated with a lack of lesion regression, with sensitivity and specificity of 0.556 and 0.800, respectively. Similarly, patients with sleep disorders showed a higher risk of a lack of lesion regression compared to those without sleep disorders, with sensitivity and specificity of 0.778 and 0.783, respectively. For the hrHPV clearance, the univariate analysis showed no statistically significant clinicopathological characteristics (Table 4).

Table 3. Univariate and multivariate regression analysis of factors influencing the effect of ALA-PDT on HSIL/CIN2 regression.

Characteristics	Univariate Analysis		Multivariate Analysis	
	p	OR (95% CI)	*p*	OR (95% CI)
age	0.806	0.987 (0.891–1.093)		
NLR	0.834	1.089 (0.489–2.427)		
PLR	0.451	1.006 (0.990–1.023)		
current smoking	0.031	5.000 (1.162–21.509)	0.046	5352.87 (1.15–2.48 × 10^7)
drinking history	0.142	3.250 (0.674–15.671)		
sleeping disorders	0.003	12.654 (2.341–68.387)	0.039	5466.59 (1.58–1.90 × 10^7)
HPV vaccination	0.998	0.000 (0.000–0.000)		
HPV16/18 infection	0.846	1.158 (0.263–5.097)		
multiple types of HPV infection	0.334	2.024 (0.484–8.453)		
TCT	0.383	1.875 (0.457–7.700)		
history of giving birth	0.703	0.723 (0.136–3.834)		
abortion history	0.840	1.167 (0.262–5.186)		
combined with CO_2 laser	0.206	2.500 (0.604–10.344)		

Table 4. Univariate regression analysis of factors influencing the effect of ALA-PDT on hrHPV clearance.

Characteristics	Univariate Analysis	
	p	OR (95% CI)
age	0.734	0.988 (0.922–1.059)
NLR	0.933	1.025 (0.575–1.827)
PLR	0.301	1.007 (0.994–1.019)
current smoking	0.104	2.531 (0.826–7.756)
drinking history	0.177	2.463 (0.666–9.105)
sleeping disorders	0.133	2.267 (0.780–6.585)
HPV vaccination	0.792	0.850 (0.254–2.843)
HPV16/18 infection included	0.116	2.407 (0.805–7.191)
multiple types of HPV infection	0.450	1.500 (0.524–4.296)
TCT	0.797	0.877 (0.323–2.380)
history of giving birth	0.532	1.412 (0.478–4.168)
abortion history	0.450	1.500 (0.524–4.296)
combined with CO_2 laser	0.976	0.984 (0.354–2.735)

4. Discussion

The application of HPV vaccines, regular screening, and timely treatment of cervical precancerous lesions are effective preventive measures for cervical cancer. Untreated HSIL/CIN2 patients have a probability of developing cervical cancer [4]. Studies indicate that the majority of patients with cervical HSIL benefit from ALA-PDT, although a small number may not respond effectively [8,9]. Therefore, evaluating factors influencing ALA-PDT efficacy in cervical HSIL patients can aid in selecting more appropriate treatments. In this study, 13 clinicopathological variables were analyzed.

Smoking has been extensively studied for its association with various diseases. Reports have shown that smoking can cause DNA strand breaks, promote viral amplification in cervical cells infected with hrHPV, and enhance the expression of HPV E6 and E7 proteins [10,11]. A meta-analysis on Japanese women demonstrated that smoking was strongly associated with cervical cancer [10]. Reports also indicate that smoking increases the risk of recurrence of genital warts and HSIL [11]. Building on this, our analysis identified current smoking as an independent risk factor for failure in lesion regression following ALA-PDT. However, current smoking did not significantly affect hrHPV clearance in HSIL/CIN2 patients after ALA-PDT. Further studies with larger sample sizes and longer follow-up periods are warranted to validate these findings.

Numerous studies have demonstrated a connection between sleep and HPV infection. Short sleep durations may increase HPV infection likelihood [12]. Enhancement of sleep quality during HPV infection is assumed to feedback to the immune system to promote host defense [13]. In particular, slow-wave sleep can support immune system recovery [14]. Further understanding of the casual relationship between sleep deprivation and immune deregulation would help to identify individuals at risk of disease and to prevent adverse health outcomes [15]. However, another study found that the quality of sleep had no bearing on HPV infection [16]. The results of this study indicated that sleep significantly influenced lesion regression following ALA-PDT but did not affect hrHPV clearance. Sleep disorders were identified as an independent risk factor for poorer lesion outcomes in HSIL/CIN2 patients treated with ALA-PDT. These findings underscore that HSIL/CIN2 patients with better sleep quality may derive greater benefit from ALA-PDT.

Our previous studies proved that age was a risk factor for ALA-PDT, with the >50 age group showing a higher lesion progression rate and lower HPV remission rate than the <50 age group [17]. However, age did not affect the outcomes of ALA-PDT in patients with HSIL/CIN2 in this study. This may be attributed to the fact that the majority of patients were younger than 50 years old (66/69, 95.65%), reflecting the preference of young women with fertility concerns opting for ALA-PDT.

Drinking habits and HPV vaccination are known clinical factors that influence HPV infection [18–21]. A history of childbirth and abortion is also believed to impact vaginal health. We were interested in how these factors might affect the efficacy of ALA-PDT in HSIL/CIN2 patients. Our results showed no significant difference between the groups that achieved success and those that did not in terms of drinking habits, HPV vaccination status, or history of childbirth or abortion. However, further analysis with larger or multi-center sample sizes is needed to derive conclusive findings. Additionally, stratified analysis based on specific childbirth or abortion histories and types of HPV vaccines warrants further research.

HPV testing and the TCT are common laboratory indicators for cervical lesions. A previous study demonstrated that HPV 16/18 positive cases can benefit most from ALA-PDT, and the HPV clearance rate in patients with multiple-type HPV infection was significantly lower than that in patients with single-type HPV infection [22]. The TCT has some degree of cervical lesion presentation and CIN/HSIL diagnostic value; however, in this study, HPV16/18 infection, multiple-type HPV infections, and TCT results were not found to be associated with the outcome of ALA-PDT treatment for HSIL/CIN2, which may be due to the variations in methods or patient populations. The NLR and PLR of blood are indicators of not only systemic inflammation but also different malignancies [23]. One study reported the NLR is an independent prognostic factor for relapse-free survival following surgical resection in CIN patients [24]. Furthermore, preoperative PLR levels could be a marker for predicting recurrence/residual disease of HSIL after LEEP [24]. We found that the NLR and PLR did not affect the outcome of ALA-PDT treatment for HSIL.

ALA-PDT combined with CO_2 laser assistance is being explored to enhance its effectiveness and improve patient tolerance. The CO_2 laser itself has therapeutic benefits. Previous clinical studies have shown improved lesion response when a CO_2 laser is used before PDT compared to ALA-PDT or laser treatment alone [25,26]. However, this study revealed that there was no significant difference between the ALA-PDT+ CO_2 laser group and the ALA-PDT group.

These results can inform ALA-PDT treatment for HSIL/CIN2 patients in the future. For patients who smoke or are in a smoking environment for a long time, it is necessary to quit smoking or consciously stay away from the smoking environment before ALA-PDT treatment and maintain this until the lesions have regressed. For patients with sleep disorders, they should be helped to find the cause, through symptomatic and etiological treatment, to improve their quality of sleep. These steps require the joint efforts of patients, families, and doctors. Having regular telephone or home follow-up appointments and the timely monitoring of the lifestyle behaviors of study participants will contribute to the rigor of future clinical studies.

The major deficiencies of this study were the small, single-center sample and short follow-up, which may have caused bias. Furthermore, analysis of the clinicopathological characteristics was not detailed enough, and hierarchical analysis may help evaluate the influences of characteristics more comprehensively.

5. Conclusions

Overall, we examined the relationship between 13 clinicopathological characteristics and the impact of ALA-PDT on HSIL/CIN2 patients. Our findings indicated that smoking and sleep disorders are independent risk factors for failure in lesion regression, while age, NLR, PLR, drinking, HPV vaccine, HPV infection type, TCT results, history of giving birth and abortion, and CO_2 laser combination therapy have no effect on lesion regression. No factors affecting hrHPV clearance were found. We suggest that careful consideration is given when evaluating ALA-PDT for patients who smoke and those with sleep disorders. To make informed treatment decisions for HSIL patients, further trials with larger sample sizes and longer follow-up periods are necessary to confirm these findings and explore additional prognostic factors for ALA-PDT.

Author Contributions: Y.W.: writing—original draft, writing—review and editing, software, and investigation; J.N.: Writing—original draft, software, and methodology; L.G.: investigation and data curation; Z.H.: investigation and data curation; Z.B.: supervision, project administration, and methodology; L.Q.: writing—review and editing, supervision, resources, project administration, methodology, data curation, and conceptualization. All authors have read and agreed to the published version of the manuscript.

Funding: This work was financially supported by the Clinical Research Innovation and Cultivation Fund of Renji Hospital, Shanghai Jiao Tong University School of Medicine (RJPY-DZX-002); the Medical and Engineering Cross Research Fund of "Star of Jiao Tong University" Program of Shanghai Jiao Tong University (YG2023ZD08); the National Natural Science Foundation of China (82072865); the Health and Family Planning Scientific Research Project of Pudong New Area Health Commission (PW2022D-03); the Natural Science Foundation of Shanghai (21ZR1438600); and the Shanghai Rising-Star Program (23QC1400100).

Institutional Review Board Statement: The study received approval from the Ethics Committee of Renji Hospital, affiliated with Shanghai Jiao Tong University School of Medicine (Committee's reference number KY2021-248-B).

Informed Consent Statement: Informed consent was obtained from all subjects involved in the study.

Data Availability Statement: The original contributions presented in the study are included in the article, further inquiries can be directed to the corresponding authors.

Conflicts of Interest: The authors declare no conflicts of interest.

References

1. Siegel, R.L.; Miller, K.D.; Fuchs, H.E.; Jemal, A. Cancer statistics, 2022. *CA Cancer J. Clin.* **2022**, *72*, 7–33. [CrossRef] [PubMed]
2. Ostör, A.G. Natural history of cervical intraepithelial neoplasia: A critical review. *Int. J. Gynecol. Pathol.* **1993**, *12*, 186–192. [CrossRef] [PubMed]

3. Tainio, K.; Athanasiou, A.; Tikkinen, K.A.O.; Aaltonen, R.; Cárdenas, J.; Hernándes Glazer-Livson, S.; Jakobsson, M.; Joronen, K.; Kiviharju, M.; Louvanto, K.; et al. Clinical course of untreated cervical intraepithelial neoplasia grade 2 under active surveillance: Systematic review and meta-analysis. *BMJ* **2018**, *360*, k499. [CrossRef] [PubMed]
4. Louvanto, K.; Aro, K.; Nedjai, B.; Bützow, R.; Jakobsson, M.; Kalliala, I.; Dillner, J.; Nieminen, P.; Lorincz, A. Methylation in Predicting Progression of Untreated High-grade Cervical Intraepithelial Neoplasia. *Clin. Infect. Dis.* **2020**, *70*, 2582–2590. [CrossRef] [PubMed]
5. Cho, A.; Park, S.; Park, S.; Kang, H.S.; Shim, S.S.; Park, C.M.; Kim, S.Y. Hemoperitoneum: A complication of loop electrosurgical excision procedure. *Obstet. Gynecol. Sci.* **2019**, *62*, 138–141. [CrossRef]
6. Pallavi, P.; Girigoswami, K.; Gowtham, P.; Harini, K.; Thirumalai, A.; Girigoswami, A. Rhodamine 6G in Oxidized Sodium Alginate Polymeric Hydrogel for Photodynamically Inactivating Cancer Cells. *Curr. Pharm. Des.* **2024**. [CrossRef] [PubMed]
7. Pallavi, P.; Harini, K.; Elboughdiri, N.; Gowtham, P.; Girigoswami, K.; Girigoswami, A. Infections associated with SARS-CoV-2 exploited via nanoformulated photodynamic therapy. *ADMET DMPK* **2023**, *11*, 513–531. [CrossRef]
8. Qiu, L.; Li, J.; Chen, F.; Wang, Y.; Wang, Y.; Wang, X.; Lv, Q.; Li, C.; Li, M.; Yang, Q.; et al. Chinese expert consensus on the clinical applications of aminolevulinic acid-based photodynamic therapy in female lower genital tract diseases (2022). *Photodiagn. Photodyn. Ther.* **2022**, *39*, 102993. [CrossRef]
9. Ozog, D.M.; Rkein, A.M.; Fabi, S.G.; Gold, M.H.; Goldman, M.P.; Lowe, N.J.; Martin, G.M.; Munavalli, G.S. Photodynamic Therapy: A Clinical Consensus Guide. *Dermatol. Surg.* **2016**, *42*, 804–827. [CrossRef]
10. Sugawara, Y.; Tsuji, I.; Mizoue, T.; Inoue, M.; Sawada, N.; Matsuo, K.; Ito, H.; Naito, M.; Nagata, C.; Kitamura, Y.; et al. Cigarette smoking and cervical cancer risk: An evaluation based on a systematic review and meta-analysis among Japanese women. *Jpn. J. Clin. Oncol.* **2019**, *49*, 77–86. [CrossRef]
11. Kaderli, R.; Schnüriger, B.; Brügger, L.E. The impact of smoking on HPV infection and the development of anogenital warts. *Int. J. Color. Dis.* **2014**, *29*, 899–908. [CrossRef]
12. Hu, H.; Wu, Y.; Zhao, M.; Liu, J.; Xie, P. Sleep duration time and human papillomavirus infection risk: The U-shaped relationship revealed by NHANES data. *PLoS ONE* **2024**, *19*, e0301212. [CrossRef] [PubMed]
13. Besedovsky, L.; Lange, T.; Haack, M. The Sleep-Immune Crosstalk in Health and Disease. *Physiol. Rev.* **2019**, *99*, 1325–1380. [CrossRef] [PubMed]
14. Majde, J.A.; Krueger, J.M. Links between the innate immune system and sleep. *J. Allergy Clin. Immunol.* **2005**, *116*, 1188–1198. [CrossRef] [PubMed]
15. Garbarino, S.; Lanteri, P.; Bragazzi, N.L.; Magnavita, N.; Scoditti, E. Role of sleep deprivation in immune-related disease risk and outcomes. *Commun. Biol.* **2021**, *4*, 1304. [CrossRef]
16. Li, Y.; Liu, M.; Huang, P.; Wang, W.; Jiang, Y.; Yang, Z.; Wang, A. The lifestyle factors of physical activity and diet balance associated with HPV infection in China: The cross-sectional study. *Front. Oncol.* **2022**, *12*, 1043937. [CrossRef]
17. Gu, L.; Cheng, M.; Hong, Z.; Di, W.; Qiu, L. The effect of local photodynamic therapy with 5-aminolevulinic acid for the treatment of cervical low-grade squamous intraepithelial lesions with high-risk HPV infection: A retrospective study. *Photodiagn. Photodyn. Ther.* **2021**, *33*, 102172. [CrossRef]
18. De Sanjosé, S.; Brotons, M.; Pavón, M.A. The natural history of human papillomavirus infection. *Best. Pract. Res. Clin. Obstet. Gynaecol.* **2018**, *47*, 2–13. [CrossRef]
19. Weiderpass, E.; Ye, W.; Tamimi, R.; Trichopolous, D.; Nyren, O.; Vainio, H.; Adami, H.O. Alcoholism and risk for cancer of the cervix uteri, vagina, and vulva. *Cancer Epidemiol. Biomark. Prev.* **2001**, *10*, 899–901.
20. Wong, M.C.S.; Vlantis, A.C.; Liang, M.; Wong, P.Y.; Ho, W.C.S.; Boon, S.S.; Leung, C.; Chan, P.K.S.; Chen, Z. Persistence and clearance of oral human papillomavirus infections: A prospective population-based cohort study. *J. Med. Virol.* **2020**, *92*, 3807–3814. [CrossRef]
21. Oyouni, A.A.A. Human papillomavirus in cancer: Infection, disease transmission, and progress in vaccines. *J. Infect. Public Health* **2023**, *16*, 626–631. [CrossRef] [PubMed]
22. Cang, W.; Gu, L.; Hong, Z.; Wu, A.; Di, W.; Qiu, L. Effectiveness of photodynamic therapy with 5-aminolevulinic acid on HPV clearance in women without cervical lesions. *Photodiagn. Photodyn. Ther.* **2021**, *34*, 102293. [CrossRef] [PubMed]
23. Fang, T.; Wang, Y.; Yin, X.; Zhai, Z.; Zhang, Y.; Yang, Y.; You, Q.; Li, Z.; Ma, Y.; Li, C.; et al. Diagnostic Sensitivity of NLR and PLR in Early Diagnosis of Gastric Cancer. *J. Immunol. Res.* **2020**, *2020*, 9146042. [CrossRef] [PubMed]
24. Chun, S.; Shin, K.; Kim, K.H.; Kim, H.Y.; Eo, W.; Lee, J.Y.; Namkung, J.; Kwon, S.H.; Koh, S.B.; Kim, H.B. The Neutrophil-Lymphocyte Ratio Predicts Recurrence of Cervical Intraepithelial Neoplasia. *J. Cancer* **2017**, *8*, 2205–2211. [CrossRef]
25. Cai, H.; Wang, Y.X.; Zheng, J.C.; Sun, P.; Yang, Z.Y.; Li, Y.L.; Liu, X.Y.; Li, Q.; Liu, W. Photodynamic therapy in combination with CO_2 laser for the treatment of Bowen's disease. *Lasers Med. Sci.* **2015**, *30*, 1505–1510. [CrossRef]
26. Lippert, J.; Smucler, R.; Vlk, M. Fractional carbon dioxide laser improves nodular basal cell carcinoma treatment with photodynamic therapy with methyl 5-aminolevulinate. *Dermatol. Surg.* **2013**, *39*, 1202–1208. [CrossRef]

biomedicines

MDPI

Article

Unraveling Microviscosity Changes Induced in Cancer Cells by Photodynamic Therapy with Targeted Genetically Encoded Photosensitizer

Liubov E. Shimolina [1], Aleksandra E. Khlynova [1], Vadim V. Elagin [1], Pavel A. Bureev [1], Petr S. Sherin [2], Marina K. Kuimova [2] and Marina V. Shirmanova [1,*]

1 Institute of Experimental Oncology and Biomedical Technologies, Privolzhsky Research Medical University, Minin and Pozharsky Square, 10/1, 603005 Nizhny Novgorod, Russia; shimolina_l@pimunn.net (L.E.S.); khlynova_a@pimunn.net (A.E.K.); elagin_v@pimunn.net (V.V.E.); bureev_p@pimunn.net (P.A.B.)

2 Department of Chemistry, Imperial College London, White City Campus, London W12 0BZ, UK; p.sherin@imperial.ac.uk (P.S.S.); m.kuimova@imperial.ac.uk (M.K.K.)

* Correspondence: shirmanovam@gmail.com

Abstract: Background: Despite the fundamental importance of cell membrane microviscosity, changes in this biophysical parameter of membranes during photodynamic therapy (PDT) have not been fully understood. Methods: In this work, changes in the microviscosity of membranes of live HeLa Kyoto tumor cells were studied during PDT with KillerRed, a genetically encoded photosensitizer, in different cellular localizations. Membrane microviscosity was visualized using fluorescence lifetime imaging microscopy (FLIM) with a viscosity-sensitive BODIPY2 rotor. Results: Depending on the localization of the phototoxic protein, different effects on membrane microviscosity were observed. With nuclear localization of KillerRed, a gradual decrease in microviscosity was detected throughout the entire observation period, while for membrane localization of KillerRed, a dramatic increase in microviscosity was observed in the first minutes after PDT, and then a significant decrease at later stages of monitoring. The obtained data on cell monolayers are in good agreement with the data obtained for 3D tumor spheroids. Conclusions: These results indicate the involvement of membrane microviscosity in the response of tumor cells to PDT, which strongly depends on the localization of reactive oxygen species attack via targeting of a genetically encoded photosensitizer.

Keywords: plasma membrane; cancer cell; microviscosity; molecular rotors; fluorescence lifetime imaging microscopy FLIM; genetically encoded photosensitizer; photodynamic therapy

1. Introduction

Photodynamic therapy (PDT) is a treatment method used in clinics to address cancerous and pre-cancerous lesions [1]. The principle of PDT is a photochemical reaction between a photosensitizer and oxygen or biomolecules initiated by the absorption light, which results in the generation of reactive oxygen species (ROS). The predominant mechanism of this reaction for most photosensitizers is energy transfer from a triplet state to molecular oxygen with the formation of singlet oxygen ($^1O_2$), referred to as type II photoreactions. Another type of photoreaction, type I, relies on electron or hydrogen atom transfer between a photosensitizer and a substrate with the formation of free radicals that further react with oxygen [2]. The type of photosensitized reaction depends on many factors, such as the nature and concentration of the sensitizer, oxygen content, and substrate reactivity. In a complex biological environment, both reactions can proceed in parallel.

Due to the high reactivity and short half-life of ROS, photodamage within a cell and the subsequent biological response strongly depend on the subcellular localization of the photosensitizer [3]. Chemical photosensitizers are typically distributed in multiple sites, including mitochondria, lysosomes, ER and the Golgi apparatus, and plasma membrane [4].

In these structures, membrane lipids are more vulnerable to oxidation during PDT due to the presence of double bonds in the fatty acid tails. Lipid oxidation causes marked changes in the physicochemical properties of membranes, such as permeability, polarity, lipid order, and microviscosity. While the effects of lipid oxidation on the membrane state are relatively well investigated in model lipid bilayers, there is very limited information about the changes in cellular membranes.

Fluorescence lifetime imaging microscopy (FLIM) using fluorescent molecular rotors has proven to be a reliable method to obtain micrometer-scale quantitative maps of viscosity in live cells and even tissues and monitor it in space and time [5]. Sensing of microviscosity with such probes is based on their ability for intramolecular twisting in the excited state, which is constrained in a more viscous environment, so that emission quantum yield and decay time increase. Among molecular rotors, BODIPY (boron-dipyrromethene) derivatives are considered highly effective viscosity probes due to their excellent photophysical properties, such as high quantum yields, narrow emission bands, a wide dynamic range of viscosity values that are measurable, and monoexponential time-resolved decays allowing easy interpretation of the data [6].

Molecular rotor detection has been used to monitor a large increase in membrane viscosity in live cells and model membranes due to lipid peroxidation in type II photoreactions [7–10]. The proposed mechanism for this change was the oxidation of unsaturated bonds in lipids by singlet oxygen at the site of irradiation and the subsequent formation of long-lived peroxide species that can travel further in lipid bilayers [7]. At the same time, a decrease in local viscosity was detected following type I photoreactions [7,8]. One of the possible mechanisms is the oxidative cleavage of C=C bonds in lipids, which results in their very loose packing and pore formation [11]. These trends were determined by using specific photosensitizers that are known to produce either radical species (e.g., methylene blue) or singlet molecular oxygen (porphyrins), and the mechanism of changes was established in model membranes that were assembled from synthetic oxidized/truncated lipids. However, the links between the localization of a photosensitizer within a cell and the type of photosensitized reactions were never explored. A significant challenge in this area is the targeted delivery of the same photosensitizer to different organelles in cells to enable a direct comparison.

We set out to study how the differential localization of the sensitizer affects ROS production and the effects. The only possible way to do this unambiguously is with a genetically encoded sensitizer. We used a genetically encoded photosensitizer, KillerRed, to induce a photodynamic reaction in a specific cellular compartment. This was followed by FLIM with molecular rotor BODIPY2 to dynamically track changes in the microviscosity of plasma membranes of cancer cells during PDT induced in different cellular localizations. Specifically, we compared the effects for the two localizations—(i) KillerRed within the plasma membranes (colocalized with BODIPY2 rotor) and (ii) within nuclei. There are not many such photosensitizers known, and among them, KillerRed is one of the most extensively studied. In addition, the choice of KillerRed was due to its spectral characteristics, which allow for multiparametric analysis with a molecular rotor that fluoresces in the green range. Another well-known genetically encoded photosensitizer, miniSOG, fluoresces in the blue region of the spectrum, which complicates its use in combination with a BODIPY molecular rotor [12].

KillerRed is a fluorescent protein of the class of green fluorescent proteins (GFPs) that possesses notable phototoxic properties, efficiently generating ROS via type I photoreactions [13,14]. The cause of its phototoxicity is the presence of a water-filled channel reaching the chromophore area from the end cap of the β-barrel and of two amino acid residues (Glu68 and Ser119) adjacent to the chromophore [14]. This combination is hypothesized to trigger photoinduced reduction in the acylimine group of the chromophore with subsequent abstraction of the electron by molecular oxygen. Although KillerRed is inferior in its phototoxicity to clinically approved porphyrin-based chemical photosensitizers, it represents a unique tool to induce a precise site of photodamage within a cell. As such, it has found applications for targeted protein inactivation, induction of oxidative stress in specified organelles, and cancer cell killing in preclinical settings [15,16]. While KillerRed has been used for PDT previously, its use was not combined with ROS monitoring in general or molecular rotor-based monitoring in particular, which presents a significantly novel result.

This study was carried out on monolayer HeLa Kyoto (human cervical cancer) cultures and in three-dimensional tumor spheroids stably expressing KillerRed within the plasma membrane or cellular nuclei. Photobleaching studies of KillerRed, cell viability, and ROS analyses were performed to validate the efficacy of PDT. The microviscosity of plasma membranes was monitored using FLIM and the fluorescent molecular rotor BODIPY2 localized in the hydrophobic lipid tail region of a lipid bilayer, which was previously used both in cell monolayers and in spheroids [17].

2. Materials and Methods

2.1. Cell Cultures and Generation of 3D Tumor Spheroids

HeLa Kyoto (human cervical cancer) cells with the phototoxic protein KillerRed fused to histone H2B (H2B) and plasma membrane (PM) were used. The cell lines stably expressing KillerRed in these localizations have previously been obtained by lentiviral transduction [18]. The mechanism of action of KillerRed as a genetically encoded photosensitizer is schematically depicted in Figure 1. The cells were cultured in DMEM (Life Technologies, Carlsbad, CA, USA) containing 100 µg/mL penicillin, 100 µg/mL streptomycin sulfate, 2 mM L-glutamine, and 10% fetal bovine serum (FBS) at 37 °C in a humidified atmosphere with 5% CO_2.

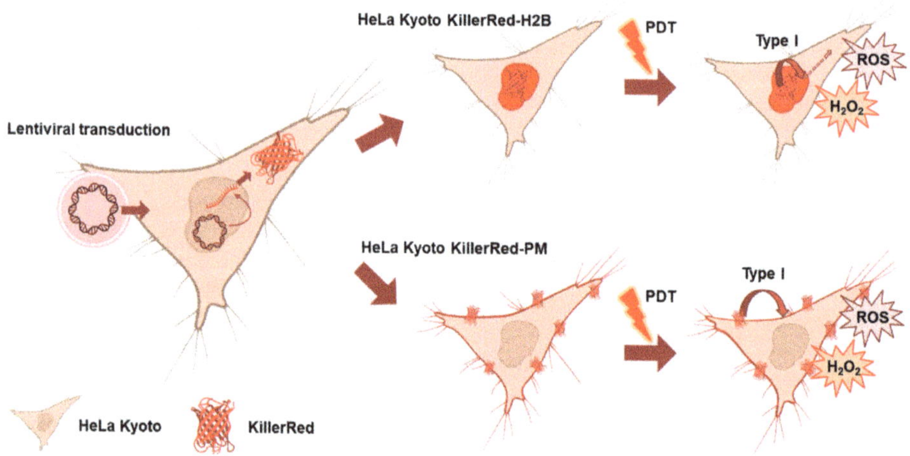

Figure 1. The schematic mechanism of action of KillerRed protein as a genetically encoded photosensitizer for PDT. The gene encoding KillerRed is incorporated in cancer cells (HeLa Kyoto) through lentiviral transduction to ensure the expression of the protein in a targeted compartment—the cell nucleus (KillerRed-H2B) or the plasma membrane (KillerRed-PM). PDT with KillerRed results in the formation of ROS via the type I photoreaction, which leads to oxidative damage in the targeting site and other compartments.

For PDT, the cells were seeded on 96-well plates with a glass bottom and black walls (SPL LifeSciences, Pocheon, Republic of Korea) in complete DMEM without phenol red (Life Technologies, Carlsbad, CA, USA). The complete medium was replaced with serum-free DMEM and irradiation was performed. A continuous laser with a wavelength of 593 nm MGL-III-593 (CNI, Qingdao, China) was used to irradiate HeLa Kyoto cells expressing the KillerRed protein. Laser radiation was directed perpendicularly to the plate surface. The laser power was controlled before each irradiation using a PM100A power meter (Thorlabs, Dortmund, Germany). The light spot diameter was 10 mm. The laser power was 50 mW/cm^2. The cells were irradiated for 25 min, which corresponds to an energy density of 75 J/cm^2.

To obtain spheroids, cells were seeded in 96-well round-bottomed, ultralow-attachment plates at a concentration of 100 cells in 200 µL of medium. Densely structured spheroids of ~300 µm in size were generated after 7 days.

For PDT, 7-day-old spheroids were carefully transferred to dark-walled 96-well plates (SPL LifeSciences, Republic of Korea) with 3–5 spheroids per well in DMEM without phenol red (Life Technologies, Carlsbad, CA, USA). PDT was performed using an MGL-III-593 laser (CNI, China) at a wavelength of 593 nm. The laser power was controlled before each irradiation using a PM100A power meter (Thorlabs, Germany). The intensity was 50 mW/cm^2, the exposure time was 25 min, and the light dose was 75 J/cm^2. The 3D spheroids were treated independently at each time point. Unirradiated spheroids served as controls.

2.2. Viscosity Imaging by FLIM with Molecular Rotor

Previously, we developed and tested protocols for staining the plasma membrane of cell monolayers and tumor spheroids [17]. The first step of staining was replacing the culture medium in the wells with ice-cold Hanks's solution without Ca^{2+}/Mg^{2+} to slow endocytosis. The cells were incubated at 4 °C for 3–5 min. Then Hanks's solution was replaced with ice-cold BODIPY2 in PBS (4.5 µM for monolayer, 8.9 µM for spheroids). FLIM images were acquired for ~30 min after staining the cells with BODIPY2. The molecular structure of molecular rotor BODIPY2 is presented in Figure 2.

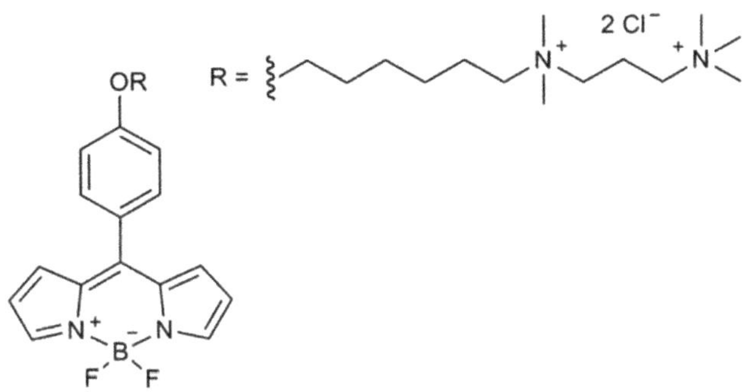

Figure 2. The molecular structure of molecular rotor BODIPY2.

For viscous imaging, a LSM 880 laser scanning microscope (Carl Zeiss, Gottingen, Germany) equipped with a FLIM SPC 150 TCSPC module (Becker & Hickl GmbH, Berlin, Germany) and a Mai Tai HP femtosecond laser (80 MHz, 140 fs, Spectra Physics, Milpitas, CA, USA) were used. The excitation wavelength of BODIPY2 fluorescence in the two-photon mode was 850 nm, and the signal was detected in a range of 500 to 550 nm using filters. A C Plan-Apochromat 40×/1.3 NA objective was used for imaging. FLIM images were acquired at a laser power of ~6–8 mW and photon collection time of 60 s to ensure ≥5000 photons per decay curve using a binning factor of 1. Ten randomly selected fields of view were collected in each well.

SPCImage 8.3 software (Becker & Hickl GmbH, Germany) was used for molecular rotor fluorescence lifetime analysis in cell membranes. Fluorescence lifetime was analyzed in the plasma membrane of cultured cells by manual selection of regions of interest. BODIPY2 fluorescence decay curves were fitted to a monoexponential decay model. The goodness of fit (χ^2 value) ranged from 0.8 to 1.2. Fluorescence lifetimes were converted to viscosity values using a calibration curve obtained previously [17].

2.3. Cell Viability Analysis

Live/dead cell counts were assessed using a live/dead dual-staining kit (Sigma, St. Louis, MO, USA). Calcein AM and propidium iodide (PI) staining was performed according to the manufacturer's protocol. Monolayer cells were stained after 10 min, 1 h and 24 h, and spheroids were stained after 3 h and 6 h. Fluorescence images were obtained using an LSM 880 laser scanning microscope (Carl Zeiss, Gottingen, Germany). The calcein fluorescence excitation wavelength was 488 nm, and detection was in the range of 500–570 nm. The PI fluorescence excitation wavelength was 543 nm, and detection was in the range of 600–700 nm. The percentage of dead cells (stained with PI) from the total number of cells was calculated for quantitative assessment of live/dead cells.

2.4. Reactive Oxygen Species Assay

For reactive oxygen species analysis, 2′,7′-dichlorofluorescein diacetate (DCFH-DA) (Sigma, USA) was used. The cells were incubated with the dye at a concentration of 1 μM for 15 min in an incubator in a darkened box. Then, they were washed with PBS and imaging was performed. The cells were stained after 10 min and 1 h. Fluorescence images were obtained using an LSM 880 laser scanning microscope (Carl Zeiss, Gottingen, Germany). Probe fluorescence was excited using an argon laser at a wavelength of 488 nm, and fluorescence was recorded in the range of 500–570 nm.

2.5. Products of Lipid Peroxidation Assay

To detect diene and triene conjugates and Schiff bases, 0.1 mL of the control sample or after PDT was taken and 5 mL of a heptane–isopropanol mixture in a ratio of 3:7 was added to it. The resulting mixture was shaken for 15 min. After that, 1 mL of 0.01 M aqueous hydrochloric acid solution was added for phase separation. After phase separation, the upper heptane phase was transferred to a separate test tube, and 1 g of sodium chloride was added to the lower one to dehydrate the isopropanol extract, which was transferred to a clean test tube. For quantitative assessment, the transmittance was measured on a spectrophotometer. Each phase was evaluated against the control at wavelengths of 220, 232, 278, and 400 nm. The content of diene and triene conjugates and Schiff bases was estimated by the ratio of the transmittances at 232 nm (for diene conjugates), 278 nm (for triene conjugates), and 400 nm (for Schiff bases) to the transmittance at 220 nm (isolated double bonds). The obtained relative content of the products is presented as percentage of the negative control.

2.6. MTT Assay

For viability analysis, cells were seeded in 96-well plates (10×10^3 cells/well) and incubated for 24 h. Then, PDT was performed. After 24 h, cells were treated with MTT (3(4,5-dimethyl-2-thiazolyl)-2,5-diphenyl-2*H*-tetrazolebromide) reagent (PanEco, Moscow, Russia) according to the manufacturer's protocol and colorimetric analysis was performed at 570 nm using a multi-mode microplate reader (Synergy Mx; BioTek Instruments, Winooski, VT, USA). Cell viability was calculated as a percentage of untreated control cells. Three independent experiments with 8–10 internal replicates were performed for each cell line.

2.7. Statistical Analysis

All experiments were performed in triplicate, the results from which were then combined. Data are expressed as means ± SD. To estimate the statistical significance of the differences, ANOVA with Bonferroni's post hoc test or a two-tailed Student's *t*-test were used where appropriate ($p < 0.05$ was considered statistically significant).

3. Results

3.1. Cell Viability and ROS After PDT with KillerRed

To induce oxidative stress in a specific cellular compartment, the genetically encoded photosensitizer KillerRed was targeted either to the plasma membrane or to the histone protein H2B in the nucleus. Figure 3 shows fluorescence signals of KillerRed in these localizations (emitting at 610 nm), overlapped with fluorescence of molecular rotor BODIPY2 (at 520 nm) localized in the cell plasma membrane. For nuclear localization (H2B) of a phototoxic protein, the rotor signal in the membrane and the photosensitizer signal in the nucleus can be clearly distinguished. For plasma membrane localization of KillerRed (PM), the signals show a significant overlap (Figure 3, merged column). The Manders colocalization coefficient was 0.842.

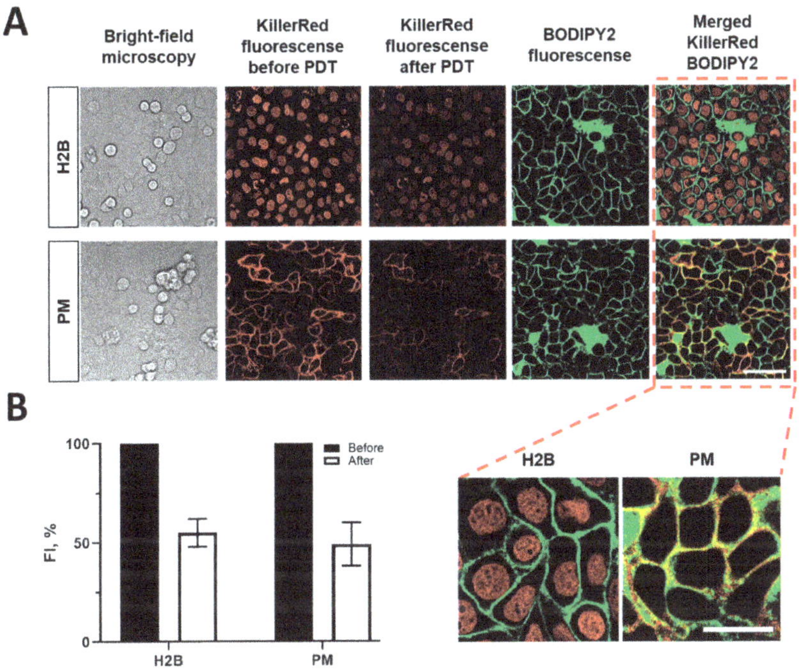

Figure 3. (**A**) Localization of the genetically encoded photosensitizer KillerRed in the nucleus and plasma membrane of HeLa cells and of the fluorescent molecular rotor in the plasma membrane. The scale bar is 40 μm, applicable to all images. (**B**) Photobleaching of KillerRed after PDT. Quantification of the fluorescence intensity in the cells after PDT. Means ± SD, $n = 50$ cells. The scale bar is 40 μm, applicable to all images. H2B: cells with nuclear localization of KillerRed, PM: cells with membrane localization of KillerRed.

Irradiation of HeLa cells expressing KillerRed at 593 nm, 50 mW/cm^2, and 75 J/cm^2 for 23 min resulted in the photobleaching of the protein by ~45% in the case of its nuclear localization and by ~55% in the case of membrane localization, indirectly indicating the occurrence of a photodynamic reaction.

Despite a similar photobleaching rate for the two localizations of KillerRed, the intracellular level of ROS detected by the DCFH-DA dye was different (Figure 4C). In the case of KillerRed-H2B, ROS production had almost doubled at 10 min after PDT and remained at the same level after 1 h. In the case of KillerRed-PM, the amount of ROS was ~3.5 times higher than in the untreated control in the time period of 10 min to 1 h. At 24 h after PDT, the amount of ROS was comparable with the control for both localizations of KillerRed.

Using this ROS detection method, we cannot decouple different rates of ROS production from different rates of local ROS reactions. Therefore, assuming KillerRed produced the same concentration of ROS under identical irradiation conditions at these two locations, one possible explanation for the higher concentration of ROS detected with DCFH-DA for KillerRed-PM is a higher concentration of oxygen within lipid bilayers [19]. A ready attack of ROS on histones and DNA, which are close to the phototoxic protein, and the presence of various antioxidant enzymes and molecules in nuclei, which can neutralize ROS, may also lead to lower ROS generation in the case of KillerRed-H2B [20,21].

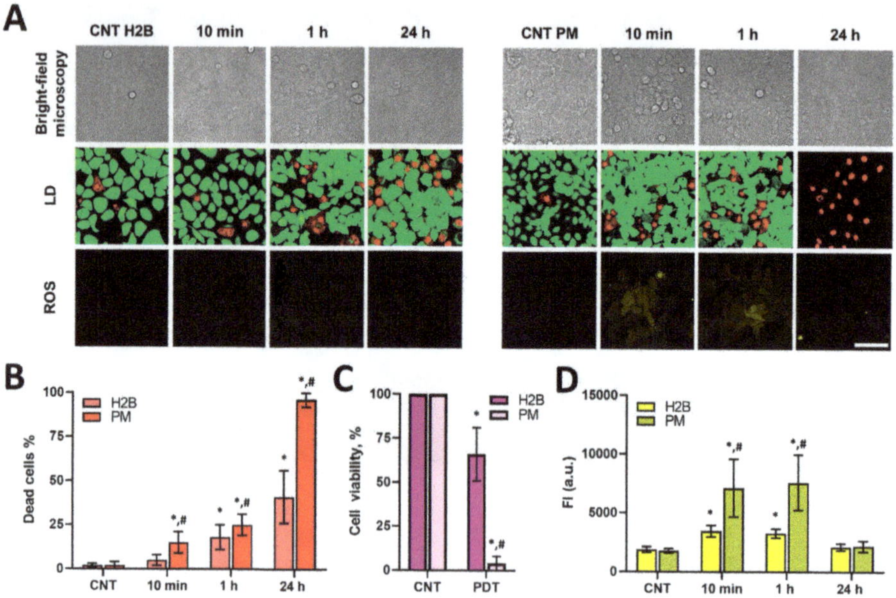

Figure 4. Cell viability and ROS analysis in cells after PDT with genetically encoded KillerRed protein. (**A**) Live (green)/dead (red) cells (LD) and ROS assay fluorescence images 10 min and 1 h after PDT. The bar is 40 μm, applicable to all images. (**B**) Quantitative analysis of dead cells in control and treated cell populations, %. (**C**) Viability of control and treated cells in 24 h after PDT, determined using the MTT assay. (**D**) Quantitative analysis of ROS in control and treated cell populations. Fluorescence intensity of DCFH-DA is shown as means ± SD. CNT: control with different localization of KillerRed. H2B: cells with nuclear localization of KillerRed. PM: cells with membrane localization of KillerRed. * $p < 0.05$ with control; # $p < 0.05$ with KillerRed-H2B.

Cell viability assays revealed that cells with membrane-localized KillerRed died faster compared to ones with nuclei-localized KillerRed (Figure 4B). In the case of KillerRed-PM, the number of dead cells increased from 2% to 15% in 10 min after irradiation and increased to 25% ($p = 0.00001$) within 1 h after PDT. After 24 h of PDT, almost all cells were dead. In the case of KillerRed-H2B, the number of dead cells reached 18% at the 1 h mark and 41% at 24 h ($p = 0.00007$). In addition, viability analysis using the MTT assay showed that the proportion of viable cells decreased to 66% after PDT with KillerRed-H2B and to 4% after PDT with KillerRed-PM.

Therefore, PDT with a photosensitizer localized within the plasma membrane was more effective than with the same photosensitizer in the chromatin at identical irradiation conditions. This was confirmed by our cell survival assays and was consistent with ROS assays.

3.2. Lipid Peroxidation After PDT with KillerRed

It is known that double bonds in unsaturated fatty acids, which are part of membrane lipids, are susceptible to oxidation, which is one of the key effects of PDT [22]. To verify that PDT with KillerRed induced lipid damage and peroxidation, its products—diene (DC) and triene (TC) conjugates—as well as Schiff bases (SB) were measured in the heptane and isopropanol phases, attributed correspondingly to the neutral and charged lipids extracted from the cells. Neutral lipids are mainly composed of triacylglycerols and wax esters, whereas polar lipids are primarily membrane glycolipids and phospholipids. The heptane phase releases neutral lipids, which are thought to make up lipid droplets and also play an important role in preventing lipotoxicity and oxidative stress. In the isopropanol phase, charged lipids are released and mainly included in the membranes, including organelle membranes.

It has been established that PDT with KillerRed at both localizations caused lipid peroxidation. In 10 min after PDT, TC and SB signals increased in the heptane phase (neutral lipids) for both localizations of KillerRed. The most pronounced effects were detected at 1 h after PDT. In the heptane phase, the DC signal increased by 11%, TC by 19%, and SB by 19% after PDT with KillerRed-H2B relative to control cells (Figure 5A). In the isopropanol phase (charged lipids), the DC signal increased by 1.5%, TC by 8.5%, and SB by 7.5% relative to control cells (Figure 5B).

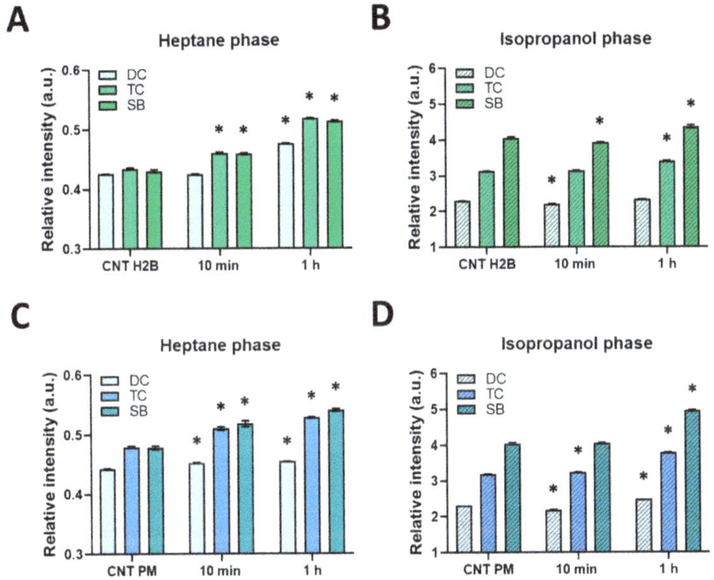

Figure 5. Analysis of products of lipid peroxidation: diene (DC) and triene (TC) conjugates and the Schiff bases (SB) for nuclear (**A**,**B**) and membrane localizations (**C**,**D**) of KillerRed. Means ± SD, $n = 5$ measurements. * Statistically significant differences in untreated control with KillerRed ($p < 0.05$). CNT H2B: control with nuclear localization of KillerRed, CNT PM: control with membrane localization of KillerRed.

PDT with KillerRed-PM resulted in an increase in the DC signal in the heptane phase by 3%, TC by 10%, and SB by 13% in 10 min irradiation (Figure 5C). The level of peroxidation products remained high at 1 h after PDT. A statistically significant increase in all three components was recorded in the isopropanol phase 1 h after PDT: the DC signal increased by 7%, TC by 19%, and SB by 23% (Figure 5D).

Interestingly, the local photosensitizer affected lipid peroxidation products. Thus, the greatest changes relative to control were recorded for neutral lipids when using KillerRed for nuclear localization and for charged lipids if peroxidation was initiated by KillerRed localized within the plasma membrane.

3.3. *Plasma Membrane Microviscosity Changes After PDT in Monolayer Cells*

The plasma membrane is the outermost boundary of each cell, and its fluidity and permeability are crucially important for rates of cell migration, ingress of chemicals, and protein interactions inside cells. We used fluorescence lifetime imaging microscopy (FLIM) in combination with a BODIPY2 molecular rotor, which was previously shown to selectively stain the plasma membrane of various cell lines [17,23]. Its lifetime can be directly correlated with microviscosity and lipid packing [24].

We recorded the FLIM of BODIPY2 Hela Kyoto cells after PDT with the phototoxic protein KillerRed localized either in the nucleus or in the plasma membrane (Figure 6A). In control cells expressing KillerRed, the fluorescence decay time of the rotor in the plasma membrane was ~2.88 ns, which corresponded to a viscosity value of 400 cP irrespective of the localization of KillerRed.

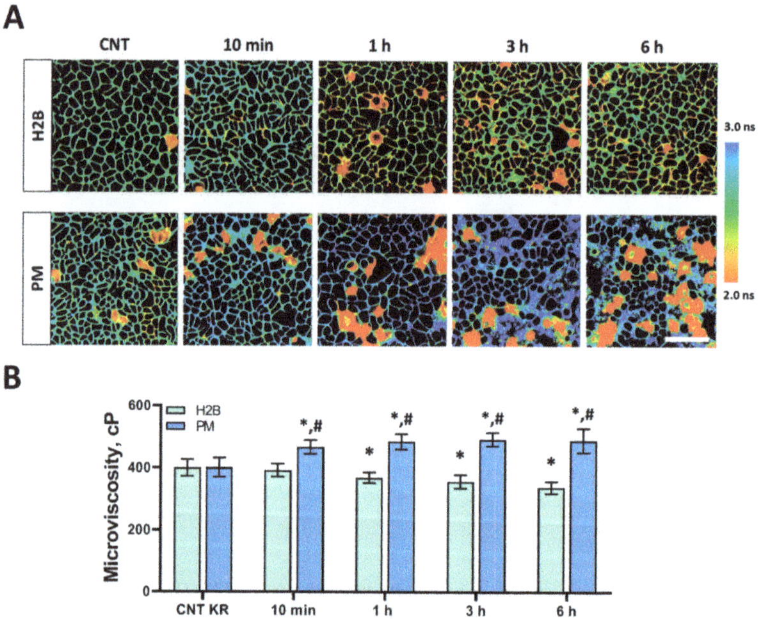

Figure 6. Plasma membrane viscosity in HeLa Kyoto cells with KillerRed during PDT. (**A**) Representative FLIM images of cells with both localizations of KillerRed. The bar is 40 μm, applicable to all images. (**B**) Quantification of viscosity of plasma membranes in HeLa Kyoto cells. Means ± SD, $n = 100$ cells for each time point. * $p < 0.05$ with control; # $p < 0.05$ with KillerRed-H2B. CNT KR: control with different localization of KillerRed. H2B: cells with nuclear localization of KillerRed. PM: cells with membrane localization of KillerRed.

Following continuous irradiation inducing PDT and cell death (Figure 3), viscosity was measured. Interestingly, the effects of PDT on the microviscosity of plasma membranes of cells were different for different localizations of the photosensitizer KillerRed (Figure 6). In cells with KillerRed-H2B, microviscosity gradually decreased to 369 ± 17 cP ($p = 0.0009$) at 1 h after PDT and then to 338 ± 19 cP ($p = 0.0006$) at 6 h.

For cells with membrane localization of the photosensitizer KillerRed-PM, a dramatic increase in membrane microviscosity was initially recorded. In the first minutes after PDT, microviscosity increased from 405 ± 31 cP to 467 ± 22 cP (p = 0.0007) and continued to increase. At 6 h post-PDT, the value was 489 ± 39 cP (p = 0.0005). However, the cells were not viable 24 h after PDT, causing BODIPY2 internalization, so microviscosity measurements were not performed in this case.

3.4. Changes in Membrane Microviscosity After PDT in 3D Spheroid Cells

Next, we examined the effects of PDT on the membrane microviscosity of three-dimensional tumor spheroids. This in vitro tumor model is more intricate than monolayer cells, as it better mimics cell–cell interactions and tumor heterogeneity. Control untreated spheroids were represented by structures with a dense core 200–250 μm in diameter.

As was the case for 2D cell cultures, we saw significantly higher phototoxicity for membrane-localized KillerRed compared to nuclear Killer Red (Figure 7B).

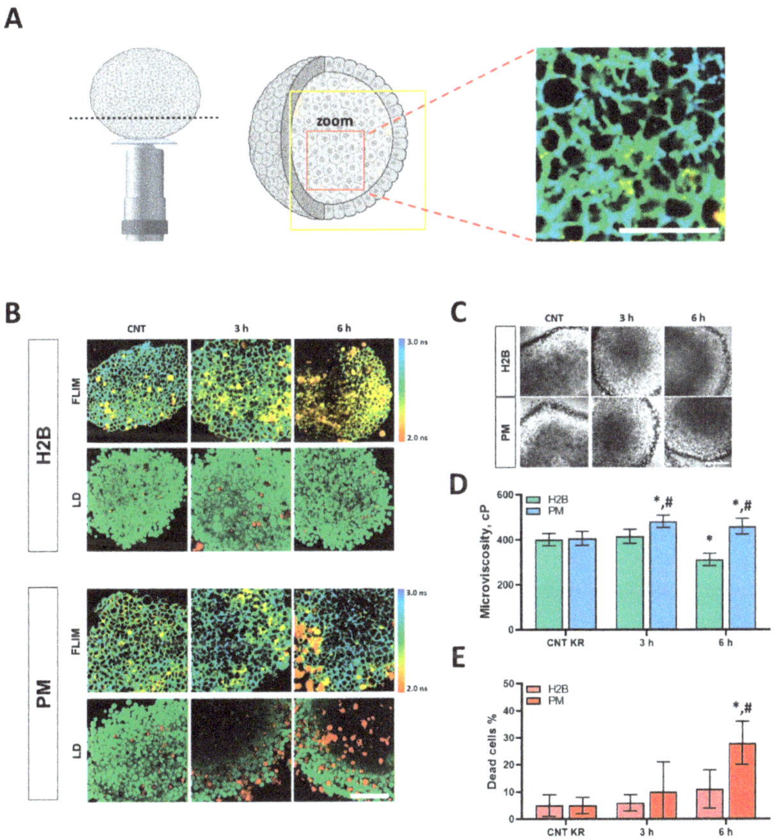

Figure 7. Plasma membrane microviscosity in HeLa tumor spheroids after PDT with KillerRed localized in the nuclei (H2B) or within the plasma membrane (PM). (**A**) Schematic representation of the spheroid area (shown by the yellow square) imaged by FLIM. The spheroid had adhered to the glass bottom, and the images were acquired from a depth of ~30 μm. Higher-magnification image of the molecular rotor distribution in spheroid cell membranes indicated by the red squares. The scale bar is 80 μm. (**B**) FLIM images and live/dead (LD) assay of control and treated cells in spheroids.

Bar = 80 μm. (**C**) Morphology of control and treated spheroids. The scale bar is 80 μm. (**D**) Quantification of membrane microviscosity of spheroid cells after PDT. Means ± SD, $n = 4$ spheroids, 60 cells in each. (**E**) Quantitative analysis of dead cells in control and treated cell populations, %. * $p < 0.05$ with control; # $p < 0.05$ with KillerRed-H2B. CNT KR: control with different localization of KillerRed. H2B: cells with nuclear localization of KillerRed. PM: cells with membrane localization of KillerRed.

After PDT with KillerRed-H2B, membrane microviscosity in tumor spheroids was lower than in the control starting from 6 h: 311 ± 28 cP ($p = 0.00009$) (Figure 7D). In addition, spheroids became loose with the disappearance of the core and the decrease in cell density in spheroids. The number of dead cells increased from 5% to 11% by 6 h after PDT (Figure 7E).

In the case of KillerRed-PM, membrane microviscosity increased dramatically from 406 ± 31 cP to 481 ± 27 cP ($p = 0.00001$) at 3 h and remained high for 6 h after PDT (Figure 7D). The number of dead cells had increased from 4% to 28% by 6 h after PDT (Figure 7E). Therefore, changes in membrane microviscosity induced by PDT with genetically encoded photosensitizer KillerRed were similar for cells in both 2D (monolayer) and 3D (tumor-like spheroids) structures, although the dynamic of changes was slower in the case of spheroids.

At this point, we can only speculate about the differential microviscosity trends observed in membrane vs. nuclear-targeted photosensitizers, because we did not collect the lipidomic data, e.g., on peroxidized products, that could confirm our hypothesis. However, according to the previous studies in model systems, decreased microviscosity of the membrane (seen upon irradiation of KillerRed in the nucleus) corresponds to radical products' production, while increased microviscosity (seen upon the use of KillerRed in the plasma membrane) indicates the production of peroxides.

4. Discussion

One of the major effects of PDT at the cellular level is the oxidation of lipids constituting the membranes. PDT-induced cell death due to destruction of membranes is mediated by alterations in their physical state, which, however, is insufficiently investigated at present.

Here, we demonstrated that the early effects of PDT on the microviscosity of plasma membranes of cancer cells depend on the subcellular localization of the photosensitizer, which has not been shown previously.

The specific targeting of the photosensitizer and thus the induction of the oxidative stress exclusively within a specific cellular compartment became possible with the use of the genetically encoded phototoxic protein KillerRed. KillerRed is a red dimeric fluorescent protein that acts as a type I photosensitizer of Type I. The ability of KillerRed to kill cancer cells upon irradiation with light, both in cellulo and in vivo, was reliably established in previous works by our group and others [14,18]. It was found earlier that KillerRed at plasma membrane causes mainly necrosis of cells [15,16,25], while KillerRed localized in chromatin induces damage of DNA, activation of reparation machinery, and blockage of cell division [26,27].

Different effects of PDT on membrane microviscosity have been described, which was mainly associated with the type of photoreaction (electron transfer or singlet oxygen-based) and consequently with different products reacting with lipids. The major mechanism for the increase in viscosity was proposed to be peroxidation of lipids by singlet oxygen following type II reactions. There are several studies that have demonstrated an increase in viscosity/decrease in fluidity of lipid membranes as a result of lipid peroxidation during PDT. For instance, Paez-Perez et al., using FLIM with molecular rotors, detected higher microviscosity in model membranes consisting of peroxidized lipids compared to non-oxidized lipids. The polar hydroperoxide functional group -OOH from peroxidized lipid increased molecular ordering within the membrane and the lateral heterogeneity of the bilayer with the appearance of more ordered lipid clusters [28]. Vyšniauskas et al. recorded an increase in viscosity of lipid bilayer in a model of giant unilamellar vesicles (GUVs) in the presence of three different porphyrin-based photosensitizers [7]. Lakos et al. observed, using fluorescence anisotropy measurements, a decrease in membrane fluidity in mouse myeloma cells and phosphatidylcholine liposomes with membrane-bound hematoporphyrin as a photosensitizer [29]. Our recent research revealed, using FLIM, that PDT with the chlorine e6-based photosensitizer Photoditazine caused increased microviscosity in plasma membrane in cultured cancer cells and tumor xenografts [30]. Other studies have reported on the increase in local microviscosity in the cell cytoplasm during PDT. For example, with the use of a porphyrin dimer-based ratiometric rotor, Kuimova et al. demonstrated significantly increased intracellular viscosity during photoinduced cell death [9]. In a study by Izquierdo et al., PDT in cells using cyano-aryl porphyrazine was accompanied by a significant viscosity increase [10]. Higher microviscosity was also observed in mitochondria during PDT with mitochondria-targeting photosensitizers [31]. In the case of membrane localization of KillerRed, we observed a pronounced increase in membrane microviscosity in monolayer cells in the period from 10 min to 6 h after PDT. Thus, our results with membrane-targeting KillerRed-PM are consistent with these studies in terms of the effect on viscosity. Interestingly, the increased microviscosity was previously associated with type II photosensitizers, i.e., a direct reaction of singlet oxygen $^{1}O_2$ with unsaturated lipids, producing peroxidation products. However, in a type I reaction, which is more probable for KillerRed, lipid hydroperoxide can also be generated with hydroxyl radical HO· as the proximal lipid oxidant [32]. Therefore, a strong interaction between a membrane-bound KillerRed and membrane lipids likely favored photoinduced peroxidation.

Decreased microviscosity of a lipid bilayer after PDT was observed for a type I oxidation process [6]. As shown in ref. [7], using methylene blue as a photosensitizer in a GUV model, decreased microviscosity of a lipid bilayer after PDT can be a result of cleavage of the double bond in lipid molecules. In addition, lipid hydroperoxides can be transformed into short-chain compounds (e.g., aldehydes) [33], the presence of which in the membrane can result in the formation of hydrophilic pores at moderate oxidative stress or in membrane disruption upon massive oxidation [34].

In our study with histone-targeting KillerRed-H2B, lower microviscosity of plasma membrane was detected starting from 1 h post-PDT in monolayer cells and from 3 h in spheroids. Since KillerRed is a type I photosensitizer, these results seem to be consistent with radical-induced cleavage. However, the principal difference of our present study is that a photosensitizer was not localized in the membrane in our experiments. It is known that ROS have an extremely short lifetime and a limited diffusion distance [35]. ROS generated by KillerRed in the nucleus can interact with different molecules in the cell before reaching the plasma membrane, giving rise to a large number of oxidation products and free radicals. Moreover, different antioxidant systems present in the cells can prevent oxidation processes through the removal of ROS [36]. Given the complexity of ROS-induced reactions in a complex biological environment, it is difficult to make any assumptions regarding the mechanisms that led to a fluidification of the plasma membrane after PDT with KillerRed-H2B.

In our previous work, we have investigated alterations of plasma membrane microviscosity induced by cytotoxic chemotherapy, which targets DNA and interferes with the cell division process [37,38], and in this sense is similar to PDT with KillerRed-H2B. However, during chemotherapy, the changes in microviscosity were more delayed (e.g., with cisplatin, oxaliplatin) or transient (e.g., with 5-fluorouracil). As such, the dynamics of viscosity changes observed after PDT with KillerRed-H2B allow us to assume that they resulted from ROS or radical attack to lipids rather than from adjustment in lipid composition in response to DNA damage.

If we compare the efficacy of PDT with KillerRed-H2B and KillerRed-PM, the former was more effective based on cell viability tests, which indicates the importance of the plasma membrane as a target for oxidative attack. According to our data, there was 2.5-fold higher percentage of cell death after PDT using membrane localization of KillerRed, and the changes in membrane microviscosity appeared earlier than with KillerRed-H2B. This means that an early increase in microviscosity of the plasma membrane due to severe ROS attack on it has more pronounced biological consequences in cancer cells than its delayed decrease mediated by nucleus damage. Although membrane fluidification has been previously associated with initiation of apoptosis via the Fas receptor [39], the efficacy of PDT with KillerRed-H2B is in general inferior to KillerRed-PM. Therefore, targeting PDT to the plasma membrane could be a more effective therapeutic strategy than targeting the nucleus.

Although oxidation of lipids is considered the major causative factor of microviscosity changes after PDT, an alternative pathway can be alterations in membrane lipid composition. Specifically, peroxidized lipids promote the formation of cholesterol domains and lipid rafts [40,41]. Cholesterol in turn is an important regulator of membrane viscosity that makes it more rigid [42]. In our previous study, lipidomic analysis revealed multiple changes in lipid profiles of cancer cells' membranes after PDT with Photoditazine: a decrease in the amount of phosphatidylcholine and monounsaturated fatty acids and an increase in cholesterol and sphingomyelin [30].

A limitation of our study is that the experiments were conducted with only one cell type—HeLa Kyoto—for which stable cell lines expressing KillerRed were available. PDT is widely used to treat cervical cancer, so the choice of HeLa cells as a model system is relevant. Typically, the patterns of PDT effects are similar in different cell types and in vivo, which we observed earlier in our studies of membrane microviscosity upon PDT and chemotherapy [30,37,38]. A limitation of in vivo measurements of viscosity using FLIM and a BODIPY2 rotor is the diffuse nature of the rotor distribution in tumor cells within the tissue. As a result, different cellular structures may contribute to the microviscosity values measured in vivo, not only the plasma membrane. Therefore, it would be difficult to correlate the viscosity data with localization of the photosensitizer. KillerRed is applicable in vivo [18,26], although its current use in vivo was outside the scope of this study. We believe that the results obtained in the framework of the present research will be reproducible in other cell types and tumors, but this requires further verification.

5. Conclusions

Damage to the lipid bilayer is an important mechanism of PDT that has multiple consequences for cell functions, including those associated with permeability and diffusion and more complex signaling cascades. Irrespective of the localization of the photosensitizer in cancer cells, the plasma membrane becomes involved in the photooxidative reactions that modify its physical state. Our study has revealed that the microviscosity of plasma membranes changes in different directions when the same photosensitizer is located in the plasma membrane or in the nucleus. Direct interaction of ROS produced by membrane-targeted KillerRed (presumably, the superoxide anion radical) with lipids in the plasma membrane caused the increase in membrane viscosity, whereas the initial production of ROS in the nucleus was followed by a decrease in viscosity of the cell membrane. Based on these results, we can conclude that targeting plasma membranes is a promising strategy to kill cancer cells by PDT. The fact that KillerRed is less phototoxic than traditional chemical photosensitizers only highlights the promise of the PDT approach aimed at local oxidative destruction of cellular plasma membranes. These findings contribute to a better understanding of the effects of oxidative stress on lipid membranes and can guide future development of novel photosensitizers. Our study is mechanistic in nature; however, it represents an important first step in improving therapeutic outcomes in the future.

Author Contributions: L.E.S.: conceptualization, methodology, investigation, formal analysis, validation, writing—original draft, visualization; A.E.K.: methodology, formal analysis, investigation; V.V.E.: investigation, methodology; P.A.B.: investigation, methodology, formal analysis; P.S.S.: investigation, formal analysis, validation, writing—original draft, visualization; M.K.K.: conceptualization, methodology, supervision, writing—original draft; M.V.S.: conceptualization, validation, writing—

original draft, funding acquisition, supervision, data curation, project administration. All authors have read and agreed to the published version of the manuscript.

Funding: The work was supported by the Russian Science Foundation (grant 20-14-00111).

Institutional Review Board Statement: Not applicable.

Informed Consent Statement: Not applicable.

Data Availability Statement: All data related to the study can be provided by the corresponding author upon reasonable request.

Acknowledgments: The authors are grateful to Konstantin Lukyanov for providing the genetically transfected cell lines Hela Kyoto-KillerRed H2B and Hela Kyoto-KillerRed mem.

Conflicts of Interest: The authors declare no conflicts of interest.

References

1. Zhao, W.; Wang, L.; Zhang, M.; Liu, Z.; Wu, C.; Pan, X.; Huang, Z.; Lu, C.; Quan, G. Photodynamic therapy for cancer: Mechanisms, photosensitizers, nanocarriers, and clinical studies. *MedComm* **2024**, *5*, e603. [CrossRef] [PubMed]
2. Baptista, M.S.; Cadet, J.; Di Mascio, P.; Ghogare, A.A.; Greer, A.; Hamblin, M.R.; Lorente, C.; Nunez, S.C.; Ribeiro, M.S.; Thomas, A.H.; et al. Type I and Type II Photosensitized Oxidation Reactions: Guidelines and Mechanistic Pathways. *Photochem. Photobiol.* **2017**, *93*, 912–919. [CrossRef] [PubMed]
3. Castano, A.P.; Demidova, T.N.; Hamblin, M.R. Mechanisms in photodynamic therapy: Part one-photosensitizers, photochemistry and cellular localization. *Photodiagn. Photodyn. Ther.* **2004**, *1*, 279–293. [CrossRef] [PubMed]
4. Kessel, D. Correlation between subcellular localization and photodynamic efficacy. *J. Porphyr. Phthalocyanines* **2004**, *8*, 1009–1014. [CrossRef]
5. Suhling, K.; Hirvonen, L.M.; Levitt, J.A.; Chung, P.-H.; Tregidgo, C.; Rusakov, D.; Zheng, K.; Ameer-Beg, S.; Poland, S.; Coelho, S.; et al. Fluorescence Lifetime Imaging. In *Handbook of Photonics for Biomedical Engineering*; Ho, A.P., Kim, D., Somekh, M., Eds.; Springer: Dordrecht, The Netherlands, 2014; pp. 1–50. [CrossRef]
6. Haidekker, M.A.; Nipper, M.; Mustafic, A.; Lichlyter, D.; Dakanali, M.; Theodorakis, E.A. Dyes with Segmental Mobility: Molecular Rotors. In *Advanced Fluorescence Reporters in Chemistry and Biology I*; Demchenko, A., Ed.; Springer Series on Fluorescence; Springer: Berlin/Heidelberg, Germany, 2010; Volume 8, pp. 267–308. [CrossRef]
7. Vyšniauskas, A.; Qurashi, M.; Kuimova, M.K. A Molecular Rotor that Measures Dynamic Changes of Lipid Bilayer Viscosity Caused by Oxidative Stress. *Chemistry* **2016**, *22*, 13210–13217. [CrossRef]
8. Sherin, P.S.; Vyšniauskas, A.; López-Duarte, I.; Ogilby, P.R.; Kuimova, M.K. Visualising UV-A light-induced damage to plasma membranes of eye lens. *J. Photochem. Photobiol. B* **2021**, *225*, 112346. [CrossRef]
9. Kuimova, M.K.; Botchway, S.W.; Parker, A.W.; Balaz, M.; Collins, H.A.; Anderson, H.L.; Suhling, K.; Ogilby, P.R. Imaging intracellular viscosity of a single cell during photoinduced cell death. *Nat. Chem.* **2009**, *1*, 69–73. [CrossRef]
10. Izquierdo, M.A.; Vyšniauskas, A.; Lermontova, S.A.; Grigoryev, I.S.; Shilyagina, N.Y.; Balalaeva, I.V.; Klapshina, L.G.; Kuimova, M.K. Dual use of porphyrazines as sensitizers and viscosity markers in photodynamic therapy. *J. Mater. Chem. B* **2015**, *3*, 1089–1096. [CrossRef]
11. Tai, W.Y.; Yang, Y.C.; Lin, H.J.; Huang, C.-P.; Cheng, Y.-L.; Chen, M.-F.; Yen, H.-L.; Liau, I. Interplay between structure and fluidity of model lipid membranes under oxidative attack. *J. Phys. Chem. B.* **2010**, *114*, 15642–15649. [CrossRef]
12. Torra, J.; Lafaye, C.; Signor, L.; Aumonier, S.; Flors, C.; Shu, X.; Nonell, S.; Gotthard, G.; Royant, A. Tailing miniSOG: Structural bases of the complex photophysics of a flavin-binding singlet oxygen photosensitizing protein. *Sci. Rep.* **2019**, *9*, 2428. [CrossRef]
13. Bulina, M.E.; Chudakov, D.M.; Britanova, O.V.; Yanushevich, Y.G.; Staroverov, D.B.; Chepurnykh, T.V.; Merzlyak, E.M.; Shkrob, M.A.; Lukyanov, S.; Lukyanov, K.A. A genetically encoded photosensitizer. *Nat. Biotechnol.* **2006**, *24*, 95–99. [CrossRef] [PubMed]
14. Pletnev, S.; Gurskaya, N.G.; Pletneva, N.V.; Lukyanov, K.A.; Chudakov, D.M.; Martynov, V.I.; Popov, V.O.; Kovalchuk, M.V.; Wlodawer, A.; Dauter, Z.; et al. Structural basis for phototoxicity of the genetically encoded photosensitizer KillerRed. *J. Biol. Chem.* **2009**, *284*, 32028–32039. [CrossRef] [PubMed]
15. Bulina, M.E.; Lukyanov, K.A.; Britanova, O.V.; Onichtchouk, D.; Lukyanov, S.; Chudakov, D.M. Chromophore-assisted light inactivation (CALI) using the phototoxic fluorescent protein KillerRed. *Nat. Protoc.* **2006**, *1*, 947–953. [CrossRef]
16. Liu, J.; Wang, F.; Qin, Y.; Feng, X. Advances in the Genetically Engineered KillerRed for Photodynamic Therapy Applications. *Int. J. Mol. Sci.* **2021**, *22*, 10130. [CrossRef] [PubMed]
17. Shirmanova, M.V.; Shimolina, L.E.; Lukina, M.M.; Zagaynova, E.V.; Kuimova, M.K. Live Cell Imaging of Viscosity in 3D Tumour Cell Models. *Adv. Exp. Med. Biol.* **2017**, *1035*, 143–153. [CrossRef] [PubMed]
18. Shirmanova, M.V.; Serebrovskaya, E.O.; Lukyanov, K.A.; Snopova, L.B.; Sirotkina, M.A.; Prodanetz, N.N.; Bugrova, M.L.; Minakova, E.A.; Turchin, I.V.; Kamensky, V.A.; et al. Phototoxic effects of fluorescent protein KillerRed on tumor cells in mice. *J. Biophotonics* **2013**, *6*, 283–290. [CrossRef]

19. Möller, M.N.; Li, Q.; Chinnaraj, M.; Cheung, H.C.; Lancaster, J.R., Jr.; Denicola, A. Solubility and diffusion of oxygen in phospholipid membranes. *Biochim. Biophys. Acta (BBA) Biomembr.* **2016**, *1858*, 2923–2930. [CrossRef]
20. Jomova, K.; Alomar, S.Y.; Alwasel, S.H.; Nepovimova, E.; Kuca, K.; Valko, M. Several lines of antioxidant defense against oxidative stress: Antioxidant enzymes, nanomaterials with multiple enzyme-mimicking activities, and low-molecular-weight antioxidants. *Arch. Toxicol.* **2024**, *98*, 1323–1367. [CrossRef]
21. Liu, H.; Zhang, J.; Zhang, S.; Yang, F.; Thacker, P.A.; Zhang, G.; Qiao, S.; Ma, X. Oral administration of Lactobacillus fermentum I5007 favors intestinal development and alters the intestinal microbiota in formula-fed piglets. *J. Agric. Food Chem.* **2014**, *62*, 860–866. [CrossRef]
22. Repetto, M.; Semprine, J.; Boveris, A. Lipid Peroxidation: Chemical Mechanism, Biological Implications and Analytical Determination [Internet]. In *Lipid Peroxidation*; InTechOpen: London, UK, 2012. [CrossRef]
23. Shimolina, L.E.; Izquierdo, M.A.; López-Duarte, I.; Bull, J.A.; Shirmanova, M.V.; Klapshina, L.G.; Zagaynova, E.V.; Kuimova, M.K. Imaging tumor microscopic viscosity in vivo using molecular rotors. *Sci. Rep.* **2017**, *7*, 41097. [CrossRef]
24. Dent, M.R.; López-Duarte, I.; Dickson, C.J.; Geoghegan, N.D.; Cooper, J.M.; Gould, I.R.; Krams, R.; Bull, J.A.; Brooks, N.J.; Kuimova, M.K. Imaging phase separation in model lipid membranes through the use of BODIPY based molecular rotors. *Phys. Chem. Chem. Phys.* **2015**, *17*, 18393–18402. [CrossRef]
25. Kobayashi, J.; Shidara, H.; Morisawa, Y.; Kawakami, M.; Tanahashi, Y.; Hotta, K.; Oka, K. A method for selective ablation of neurons in C. elegans using the phototoxic fluorescent protein, KillerRed. *Neurosci. Lett.* **2013**, *548*, 261–264. [CrossRef] [PubMed]
26. Serebrovskaya, E.O.; Gorodnicheva, T.V.; Ermakova, G.V.; Solovieva, E.A.; Sharonov, G.V.; Zagaynova, E.V.; Chudakov, D.M.; Lukyanov, S.; Zaraisky, A.G.; Lukyanov, K.A. Light-induced blockage of cell division with a chromatin-targeted phototoxic fluorescent protein. *Biochem. J.* **2011**, *435*, 65–71. [CrossRef] [PubMed]
27. Waldeck, W.; Mueller, G.; Wiessler, M.; Tóth, K.; Braun, K. Positioning effects of KillerRed inside of cells correlate with DNA strand breaks after activation with visible light. *Int. J. Med. Sci.* **2011**, *8*, 97–105. [CrossRef] [PubMed]
28. Paez-Perez, M.; Vyšniauskas, A.; López-Duarte, I.; Lafarge, E.J.; De Castro, R.L.-R.; Marques, C.M.; Schroder, A.P.; Muller, P.; Lorenz, C.D.; Brooks, N.J.; et al. Directly imaging emergence of phase separation in peroxidized lipid membranes. *Commun. Chem.* **2023**, *6*, 15. [CrossRef]
29. Lakos, Z.; Berki, T. Effect of haematoporphyrin-induced photosensitization on lipid membranes. *J. Photochem. Photobiol. B* **1995**, *29*, 185–191. [CrossRef]
30. Shimolina, L.E.; Khlynova, A.E.; Gulin, A.A.; Elagin, V.V.; Gubina, M.V.; Bureev, P.A.; Sherin, P.S.; Kuimova, M.K.; Shirmanova, M.V. Photodynamic therapy with Photoditazine increases microviscosity of cancer cells membrane in cellulo and in vivo. *J. Photochem. Photobiol. B* **2024**, *259*, 113007. [CrossRef]
31. Yang, L.; Chen, Q.; Wan, Y.; Gan, S.; Li, S.; Lee, C.-S.; Jiang, Y.; Zhang, H.; Sun, H. A NIR molecular rotor photosensitizer for efficient PDT and synchronous mitochondrial viscosity imaging. *Chem. Commun.* **2022**, *58*, 9425–9428. [CrossRef]
32. Girotti, A.W. Photosensitized oxidation of membrane lipids: Reaction pathways, cytotoxic effects, and cytoprotective mechanisms. *J. Photochem. Photobiol. B* **2001**, *63*, 103–113. [CrossRef]
33. Itri, R.; Junqueira, H.C.; Mertins, O.; Baptista, M.S. Membrane changes under oxidative stress: The impact of oxidized lipids. *Biophys. Rev.* **2014**, *6*, 47–61. [CrossRef]
34. Vernier, P.T.; Levine, Z.A.; Wu, Y.H.; Joubert, V.; Ziegler, M.J.; Mir, L.M.; Tieleman, D.P. Electroporating fields target oxidatively damaged areas in the cell membrane. *PLoS ONE* **2009**, *4*, e7966. [CrossRef] [PubMed]
35. Yuan, J.; Peng, R.; Su, D.; Zhang, X.; Zhao, H.; Zhuang, X.; Chen, M.; Zhang, X.; Yuan, L. Cell membranes targeted unimolecular prodrug for programmatic photodynamic-chemo therapy. *Theranostics* **2021**, *11*, 3502–3511. [CrossRef] [PubMed]
36. Kwon, D.H.; Cha, H.J.; Lee, H.; Hong, S.-H.; Park, C.; Park, S.-H.; Kim, G.-Y.; Kim, S.; Kim, H.-S.; Hwang, H.-J.; et al. Protective Effect of Glutathione against Oxidative Stress-induced Cytotoxicity in RAW 264.7 Macrophages through Activating the Nuclear Factor Erythroid 2-Related Factor-2/Heme Oxygenase-1 Pathway. *Antioxidants* **2019**, *8*, 82. [CrossRef] [PubMed]
37. Shimolina, L.E.; Gulin, A.A.; Paez-Perez, M.; López-Duarte, I.; Druzhkova, I.N.; Lukina, M.M.; Gubina, M.V.; Brooks, N.J.; Zagaynova, E.V.; Kuimova, M.K.; et al. Mapping cisplatin-induced viscosity alterations in cancer cells using molecular rotor and fluorescence lifetime imaging microscopy. *J. Biomed. Opt.* **2020**, *25*, 126004. [CrossRef]
38. Shimolina, L.; Gulin, A.; Khlynova, A.; Ignatova, N.; Druzhkova, I.; Gubina, M.; Zagaynova, E.; Kuimova, M.K.; Shirmanova, M. Development of resistance to 5-fluorouracil affects membrane viscosity and lipid composition of cancer cells. *Methods Appl. Fluoresc.* **2022**, *10*, 044008. [CrossRef]
39. Rebillard, A.; Tekpli, X.; Meurette, O.; Sergent, O.; LeMoigne-Muller, G.; Vernhet, L.; Gorria, M.; Chevanne, M.; Christmann, M.; Kaina, B.; et al. Cisplatin-induced apoptosis involves membrane fluidification via inhibition of NHE1 in human colon cancer cells. *Cancer Res.* **2007**, *67*, 7865–7874. [CrossRef]
40. Jacob, R.F.; Mason, R.P. Lipid peroxidation induces cholesterol domain formation in model membranes. *J. Biol. Chem.* **2005**, *280*, 39380–39387. [CrossRef] [PubMed]

41. Ayuyan, A.G.; Cohen, F.S. Lipid peroxides promote large rafts: Effects of excitation of probes in fluorescence microscopy and electrochemical reactions during vesicle formation. *Biophys. J.* **2006**, *91*, 2172–2183. [CrossRef]
42. Chakraborty, S.; Doktorova, M.; Molugu, T.R.; Heberle, F.A.; Scott, H.L.; Dzikovski, B.; Nagao, M.; Stingaciu, L.-R.; Standaert, R.F.; Barrera, F.N.; et al. How cholesterol stiffens unsaturated lipid membranes. *Proc. Natl. Acad. Sci. USA* **2020**, *117*, 21896–21905. [CrossRef]

biomedicines

MDPI

Article

Laser Emission at 675 nm: Molecular Counteraction of the Aging Process

Lorenzo Notari [1,†], Laura Pieri [2,†], Francesca Cialdai [1], Irene Fusco [2,*], Chiara Risaliti [1], Francesca Madeddu [2], Stefano Bacci [3], Tiziano Zingoni [2] and Monica Monici [1]

1 ASA Campus Joint Laboratory, ASA Research Division, Department of Experimental and Clinical Biomedical Sciences "Mario Serio", University of Florence, 50139 Florence, Italy; lorenzo.notari@unifi.it (L.N.); francesca.cialdai@unifi.it (F.C.); chiara.risaliti@unifi.it (C.R.); monica.monici@unifi.it (M.M.)

2 El.En. Group, 50041 Calenzano, Italy; l.pieri@deka.it (L.P.); f.madeddu@elen.it (F.M.); t.zingoni@elen.it (T.Z.)

3 Research Unit of Histology and Embryology, Department of Biology, University of Florence, 50121 Florence, Italy; stefano.bacci@unifi.it

* Correspondence: i.fusco@deka.it

† These authors contributed equally to the work.

Abstract: Background/Objectives: Many lasers applied in skin rejuvenation protocols show emissions with wavelengths falling in the red or near-infrared (NIR) bands. To obtain further in vitro data on the potential therapeutic benefits regarding rejuvenation, we employed a 675 nm laser wavelength on cultured human dermal fibroblasts to understand the mechanisms involved in the skin rejuvenation process's signaling pathways by analyzing cytoskeletal proteins, extracellular matrix (ECM) components, and membrane integrins. Methods: Normal human dermal fibroblasts (NHDFs) were irradiated with a 675 nm laser 24 h after seeding, and immunofluorescence microscopy and Western blotting were applied. Results: The results demonstrate that the laser treatment induces significant changes in human dermal fibroblasts, affecting cytoskeleton organization and the production and reorganization of ECM molecules. The cell response to the treatment appears to predominantly involve paxillin-mediated signaling pathways. Conclusions: These changes suggest that laser treatment can potentially improve the structure and function of skin tissue, with interesting implications for treating skin aging.

Keywords: 675 nm laser; extracellular matrix molecules; skin rejuvenation

1. Introduction

Energy-based devices and lasers are becoming increasingly common in the esthetic field, particularly for promoting rejuvenation; however, there are still gaps in users' knowledge regarding the application of these technologies. To choose safe and effective laser treatments, it is important to know how lasers interact with tissues and cells and to understand the cellular and molecular mechanisms underlying the desired esthetic effects. Laser radiation is absorbed by chromophores present in cells and tissues. Stimulating these chromophores induces a series of biological responses that can increase the mitochondrial respiratory chain's efficiency and release biochemical factors, such as growth factors and inflammatory mediators [1]. The mitochondrial respiratory chain's cytochrome c oxidase (CCO), heme proteins, porphyrins, water, and melanin are among the best known and most studied chromophores [2]. The interaction between laser radiation and biological tissues can be photochemical, photothermal, or photomechanical depending on many parameters such as the emission wavelength, the tissue's optical properties, fluence, exposure time, etc. Many lasers applied in skin rejuvenation protocols show emissions with wavelengths falling in the red or near-infrared (NIR) bands. These wavelengths can be absorbed by intracellular chromophores, such as water and CCO, affecting a wide range of biological processes in living cells and tissues, such as cytoskeleton reorganization, metabolic rearrangements, alterations in gene expression, extracellular matrix (ECM) turnover, differentiation, and proliferation [3–5].

As we get older, the ultraviolet (UV) component of sunlight has harmful effects on our skin over time that lead to enhanced matrix metalloproteinase (MMP) activity and collagen breakdown. Indeed, one of the main factors contributing to premature skin aging is photoaging caused by ultraviolet (UV) irradiation [6]. Deep wrinkles and texture changes are among the many alterations in the skin due to photoaging caused by increased MMP [7–9] and production of reactive oxygen species (ROS), which intensify oxidative stress [10]. This results in the overexpression of proinflammatory cytokines. These cytokines deteriorate the extracellular matrix (ECM), break down collagen, and form aberrant elastic fibers.

Many products and devices have been proposed to prevent, delay, and treat skin aging before cosmetic surgery is required. Among these methods, non-ablative laser treatments have been demonstrated to improve wrinkles and skin texture [11–13].

Collagen fragmentation induced by MMP may facilitate the synthesis of new collagen; indeed, a significant aspect of dermal remodeling is the deposition of new dermal collagen, which could partially explain the improvement observed following non-ablative resurfacing [12,14].

According to clinical evidence, a 675 nm wavelength can treat skin textures, scars, wrinkles, and pigmentations [15–20].

It was discovered that wrinkles and early aging of facial skin can be caused by variations in the normal collagen type I/III ratio. Magni and colleagues demonstrated that red light-based therapies can slow down or even stop the skin's aging process by promoting collagen III synthesis in human adult fibroblasts in vitro [12].

Fibroblasts are the major cell population in connective tissue. They synthesize the extracellular matrix (which includes proteoglycans, fibronectin, laminins, glycosaminoglycans, metalloproteinases, prostaglandins, and collagen types I, III, and IV), thereby creating the stroma, the structural framework of animal tissues. Membrane integrins link fibroblast cytoskeletons to the ECM. Fibroblasts contribute to injury responses and repair tissue damage in both the initiation and resolution phases.

To obtain further in vitro data on the potential therapeutic benefits in rejuvenation, we employed a 675 nm laser wavelength on cultured human dermal fibroblasts to understand the mechanisms involved in the skin rejuvenation process's signaling pathways by analyzing cytoskeletal proteins, ECM components, and membrane integrins. Changes in the expression and organization of ECM proteins are often associated with aging-related alterations in tissue structure and function [21].

By examining these proteins, we aimed to understand whether and how laser treatment could counteract the aging process at the molecular level, potentially leading to applications that can mitigate aging effects.

2. Materials and Methods

2.1. Experimental Model and Design

Normal human dermal fibroblasts (NHDFs) were used as the experimental model for this study. For their culture and maintenance, Dulbecco's Modified Eagle Medium (DMEM) supplemented with 100 μg/mL of streptomycin, 100 U/mL of penicillin, 2 mM of glutamine, and 10% fetal bovine serum (FBS) was used. Cells were incubated at 37 °C and 5% CO_2 and cultured in 24-well plates. In each experiment, two groups of samples were used: laser-treated fibroblast cultures and untreated controls. Each group included 6 samples. For samples designated for laser treatment, six wells were chosen in the well plate, ensuring at least one well distance between each treated well to prevent overlap of laser irradiation. The wells surrounding the treated samples were filled with cardboard to shield them from laser radiation. In each well, a coverslip (13 mm Ø) was placed prior to seeding to allow for cell adherence. A total of 25 × 103 cells were seeded per well. For the negative control samples (not receiving laser treatment), the same number of cells was seeded in six wells of a different plate.

2.2. Laser Source and Laser Treatment

This step only applies to cells selected for laser treatment. The laser source used in this study was a RedTouch laser (Deka Mela, Florence, Italy). Red light (675 nm) is emitted by the RedTouch laser, which uses a scanning system of 15 × 15 mm that is able to cause targeted thermal damage to the skin at a depth of 500 µm on average, reaching the dermis. Indeed, thanks to an integrated skin cooling system, it is possible to create fractional micro-zones with a 0.7 mm width (DOT) of either sub-ablative or selective thermal damage on the skin while maintaining the epidermal layer, protecting skin health and reducing downtime.

DOT pulses separated by untreated tissue heat up the region, causing collagen fibers to denature and produce new collagen [16].

For the experiment, the cells were treated using the following parameters 24 h after seeding:

- Scan time: 150 ms;
- Irradiation time: ~18 s
- DOT spacing: 500 µm;
- Power: 10 W;
- Fluence: 389.96 J/cm^2.

In a previous study, the chosen fluence was shown to be effective in inducing collagenogenesis [12].

2.3. Immunofluorescence

Twenty-four hours post-treatment, the cells were fixed for 5 min in cold acetone and then washed in phosphate-buffered saline (PBS). To block nonspecific binding sites, samples were incubated with a PBS solution containing 3% bovine serum albumin (BSA) (Sigma-Aldrich, St. Louis, MO, USA) for 1 h at room temperature. Subsequently, cells were incubated overnight at 4 °C with specific primary antibodies for phalloidin, paxillin, integrin α5β1, α-SMA, vimentin, vinculin, collagen I, collagen III, laminin, fibronectin, and matrix metalloproteinase 1 (MMP1). These markers with different functions were chosen to obtain information on different mechanisms underlying fibroblast response to the laser treatment. The cells were washed three times with PBS and then incubated with fluorescein isothiocyanate (FITC)-conjugated secondary antibodies. The samples were washed again, mounted on glass slides using Fluoromount™ aqueous mounting medium, and evaluated using an epifluorescence microscope (Nikon, Florence, Italy) at 100× magnification and imaged with a HiRes IV digital CCD camera (DTA, Pisa, Italy).

2.4. Western Blot

For protein extraction, 50 µL of lysis solution, composed of Lysis Buffer 10× (Thermo Fisher, Waltham, MA, USA) and PMSF 100× (Thermo Fisher) in milliQ water, was added to the cell pellets. The cell pellets were then subjected to mechanical disruption by vortexing.

Protein quantification was performed using the Micro BCA™ Protein Assay Kit (Thermo Fisher, Scientific, Waltham, MA, USA), which contains a "blank" reagent (reagent A), a photosensitive reagent (reagent B), and pre-made BSA standards at 2000 µg/mL. A standard curve was generated to determine the protein concentration of the samples based on absorbance, with each point on the curve consisting of a triplicate. The curve points were observed at the following concentrations: 100 µg/mL, 80 µg/mL, 60 µg/mL, 40 µg/mL, 30 µg/mL, 20 µg/mL, and 10 µg/mL. In each well, 150 µL of solution was added with the following composition:

- Curve points: 148 µL of solution at the desired concentration + 2 µL of Lysis Buffer;
- Blank: 148 µL of H_2O + 2 µL of Lysis Buffer;
- Samples: 148 µL of H_2O + 2 µL of sample.

A solution with reagent A, reagent B, and reagent C in a 25:24:1 ratio was prepared. An amount of 150 μL of the solution composed of the three reagents was added to each well. The plates were then incubated for approximately 1 h at 37 °C. Finally, the absorbance values were read using a spectrophotometer at a wavelength of 490 nm. The absorbance values were used to determine the concentrations of the samples.

All the samples were prepared for the electrophoresis run as follows:

- Sample buffer 4× (Bio-Rad, Hercules, CA, USA);
- β-mercaptoethanol 10× (Bio-Rad);
- An amount of 30 μg of protein;
- MilliQ to bring each sample to the same volume.

The samples were boiled for 5 min at 95 °C and then transferred to ice for another 5 min. Finally, they were centrifuged before loading into the wells.

For the electrophoresis run and blotting, after setting up the Mini-PROTEAN® Tetra Vertical Electrophoresis Cell (Bio-Rad) with the 4–20% MP TGX pre-cast Stain-Free Gel (Bio-Rad), both the outer and inner chambers were filled with cold Running Buffer (Bio-Rad). The properly prepared samples were then loaded into the gel wells, along with pre-stained protein standards, and a current of 300 V and 20–30 mA were applied. After the electrophoresis run, proteins were transferred to nitrocellulose membranes (Bio-Rad) using the Trans-Blot Turbo Transfer System (Bio-Rad). The membranes were cut according to the protein molecular weight prior to hybridization with specific antibodies. The cut membranes were blocked for 5 min with EveryBlot Blocking Buffer (Bio-Rad) and then probed with the following primary antibodies overnight at 4 °C: anti α-SMA, 1:1000 (Millipore); anti-vimentin, 1:1000; anti-tubulin, 1:1000 (Millipore, Billerica, MA, USA); anti-vinculin, 1:1000 (Millipore); anti-paxillin, 1:1000 (Millipore); anti-collagen III, 1:1000 (Millipore); anti-fibronectin, 1:1000 (Millipore); and anti-MMP1, 1:1000 (Millipore). This was followed by incubation with peroxidase-conjugated secondary IgGs (1:3000). Image acquisition and analysis were performed using Image Lab 6.1 software on a ChemiDoc™ Touch instrument (Bio-Rad), utilizing the fluorescence emission of protein bands separated on stain-free gels for total lane normalization. * $p < 0.05$ was considered statistically significant.

3. Results

To reveal any morphological and functional changes induced in NHDFs by the laser treatment, an integrated approach was chosen: immunofluorescence microscopy and Western blotting were applied, allowing us to obtain qualitative and quantitative information, respectively. The parameters under study were cytoskeleton components, proteins involved in cell–ECM adhesion, and ECM molecules.

Vimentin is a cytoskeletal protein that, in addition to being an important component of the intermediate filaments, helps maintain cell shape and structural integrity [22]. An immunofluorescence analysis of vimentin expression in the laser-treated NHDFs showed a denser network of cytoskeleton intermediate filaments than in the controls, suggesting a reorganization of the intermediate filaments in response to laser treatment (Figure 1 upper panels). From a quantitative point of view, the Western blot analysis did not show significant changes ($p > 0.05$) (Figure S1).

Actin filaments were stained with phalloidin, a protein that binds to actin filaments and is therefore used in immunofluorescence techniques to visualize these cytoskeletal structures. In laser-treated NHDFs, the actin filaments appeared more spread out but less thick when compared to the controls, indicating changes in actin polymerization and rearrangement of the actin filament network (middle panels in Figure 1).

α-SMA (α-smooth muscle actin), another member of the actin family, is a contractile protein involved in the contractile apparatus of smooth muscle tissue. It is also responsible for myofibroblast contraction needed for wound closure and tissue remodeling [23]. For this reason, it is considered a marker of fibroblast–myofibroblast transdifferentiation. Both immunofluorescence imaging (lower panels in Figure 1) and the Western blot analysis

(Figure S2) of NHDFs exposed to laser treatment and untreated NHDFs demonstrated that the protein distribution and expression did not change significantly.

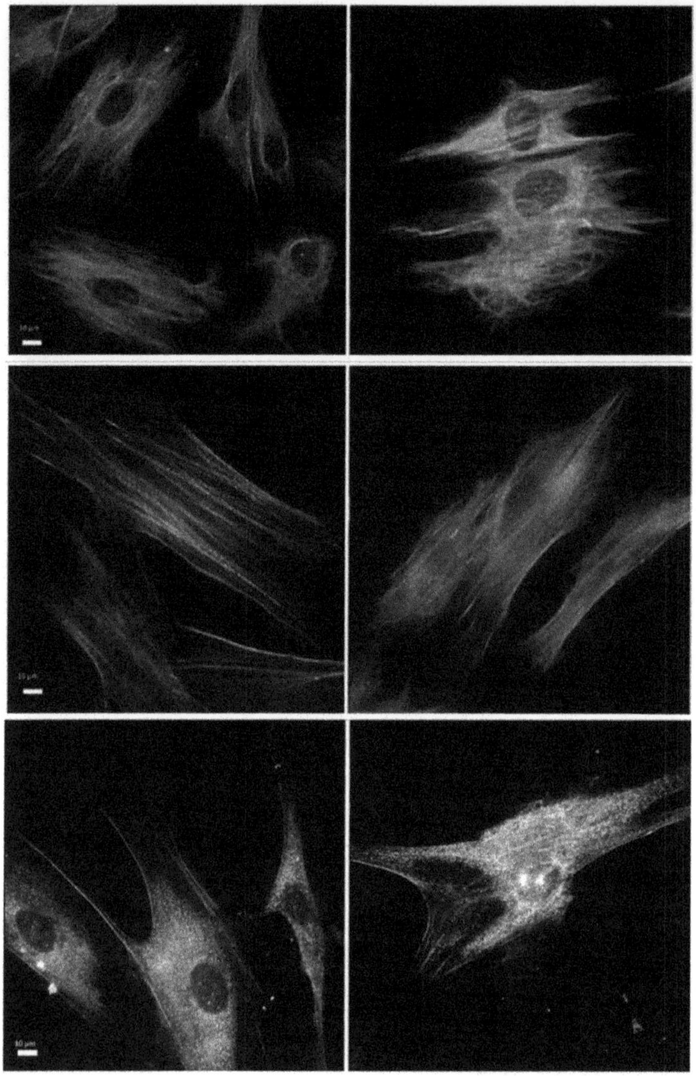

Figure 1. Effects of laser treatment on vimentin (**upper panels**), F-actin (phalloidin) (**middle panels**), and alpha-SMA (**lower panels**) using immunofluorescence analysis. Left: control samples; right: laser-treated samples.

Vinculin, a protein associated with focal adhesion junctions that links the actin cytoskeleton to the cell membrane, did not show significant changes in its distribution when analyzed by immunofluorescence microscopy in the laser-treated NHDFs compared to the controls (Figure 2). However, the Western blot analysis demonstrated a significant reduction in protein in the laser-treated samples ($p = 0.001$) (Figure 2).

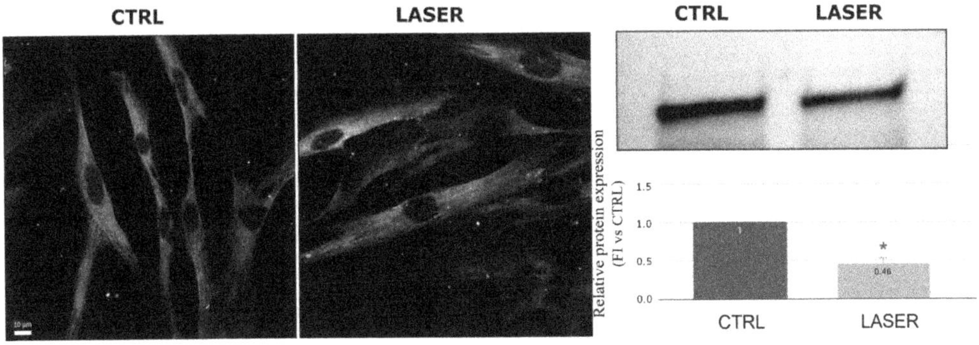

Figure 2. Effects of laser treatment on vinculin using immunofluorescence and Western blot analysis. Left: control samples; right: laser-treated samples. Western blot bands of control vs. laser-treated samples and graph showing the bands' signal intensities ($p = 0.001$). The used symbol of the asterisk (*) in the figure indicates statistically significant results compared to the control, with $p < 0.05$.

Paxillin, also involved in the formation of focal adhesions and regulating actin cytoskeleton assembly and signal transduction, showed increased expression and higher spreading in the laser-treated samples imaged using immunofluorescence microscopy (Figure 3). The Western blot analysis confirmed a significant increase in paxillin levels ($p < 0.05$) in the treated samples and revealed two bands at very close molecular weights, suggesting the presence of different paxillin isoforms. Both bands showed an increase consistent with the immunofluorescence observations (Figure 3).

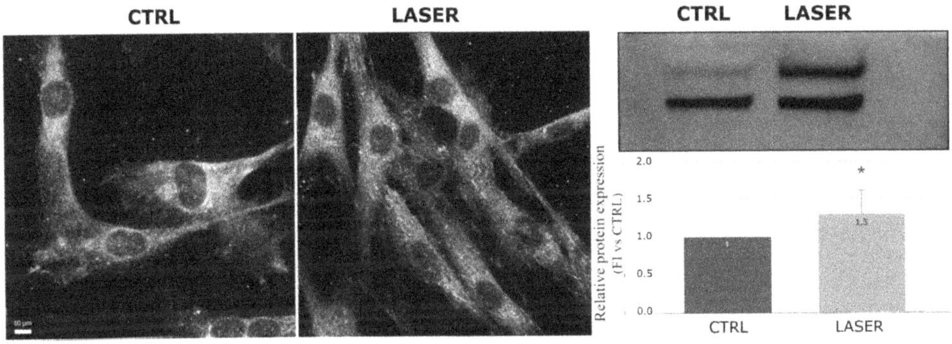

Figure 3. Effects of laser treatment on paxillin using immunofluorescence and Western blot analysis. Left: control samples; right: laser-treated samples. Western blot bands of control vs. laser-treated samples and graph showing the bands' signal intensities ($p = 0.045$). The used symbol of the asterisk (*) in the figure indicates statistically significant results compared to the control, with $p < 0.05$.

Collagen type I, the most abundant protein in the ECM, provides tensile stiffness and strength to connective tissues. Collagen I and the other collagen types are mostly produced by fibroblasts. In the laser-treated samples, immunofluorescence imaging and the fluorescence intensity analysis revealed decreased collagen I expression compared with the untreated controls (Figure S3).

On the contrary, collagen type III, a fibrillar protein that provides structural support and elasticity to tissues, including skin, showed increased expression and spreading in the cytoplasm of the laser-treated NHDFs imaged with immunofluorescence compared with the controls. The Western blot analysis also revealed a statistically significant increase ($p = 0.008$) in this protein in the laser-treated samples (Figure 4). Collagen III is also important in the response to cellular damage because it is produced in the early stages of the repair process and is crucial for the fibrillogenesis of collagen I, which increases in the late phases of repair [24].

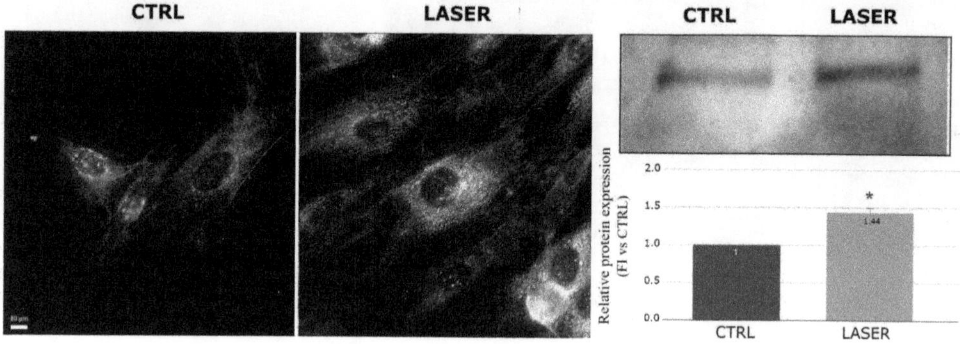

Figure 4. Effects of laser treatment on collagen III using immunofluorescence and Western blot analysis. Left: control samples; right: laser-treated samples. Western blot bands of control vs. laser-treated samples and graph showing the bands' signal intensities ($p = 0.008$). The used symbol of the asterisk (*) in the figure indicates statistically significant results compared to the control, with $p < 0.05$.

α5β1 integrin, which mediates cell adhesion to fibronectin in the ECM and plays a role in cell migration and survival, and laminin, a basal membrane protein that promotes cell adhesion, differentiation, and migration, did not show significant changes when comparing treated and untreated samples. However, in immunofluorescence imaging, the α5β1 integrin signal at the focal adhesion level appeared clearer in the laser-treated NHDFs (Figure 5). Fibrils of fibronectin, an ECM glycoprotein that promotes cell adhesion and migration and is also involved in collagen fiber formation, generated a network surrounding the cells both in the laser-treated and untreated fibroblasts. The Western blot analysis documented that fibronectin expression was significantly higher in the laser-treated fibroblasts than in the untreated ones ($p < 0.05$) (Figure 5).

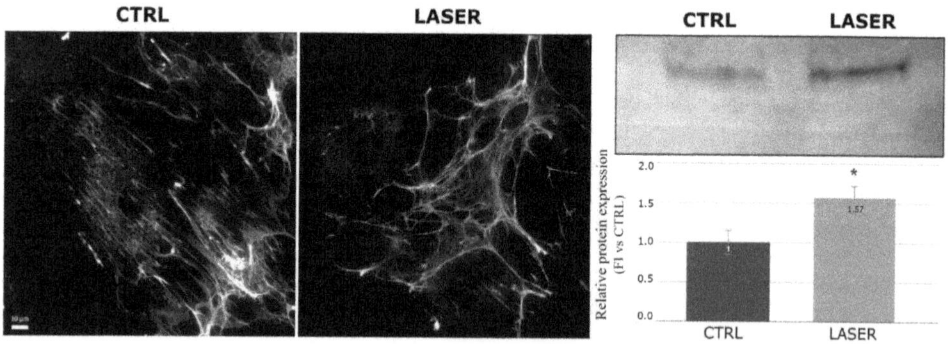

Figure 5. Effects of laser treatment on fibronectin using immunofluorescence and Western blot analysis. Left: control samples; right: laser-treated samples. Western blot bands of control vs. laser-treated samples and graph showing the bands' signal intensities ($p = 0.034$). The used symbol of the asterisk (*) in the figure indicates statistically significant results compared to the control, with $p < 0.05$.

In addition to the production and assembly of matrix molecules, such as the various types of collagen, ECM turnover is regulated by the activity of metalloproteinases, which degrade matrix components, particularly collagen [25]. MMP-1, an enzyme that degrades type I, II, and III collagen and facilitates ECM remodeling, showed higher expression in the laser-treated samples than in the controls. However, the Western blot analysis did not confirm these results. Specifically, a quantitative increase in MMP-1 was observed, but it did not reach statistical significance, probably due to the wide dispersion of the data ($p > 0.05$) (Figure 6).

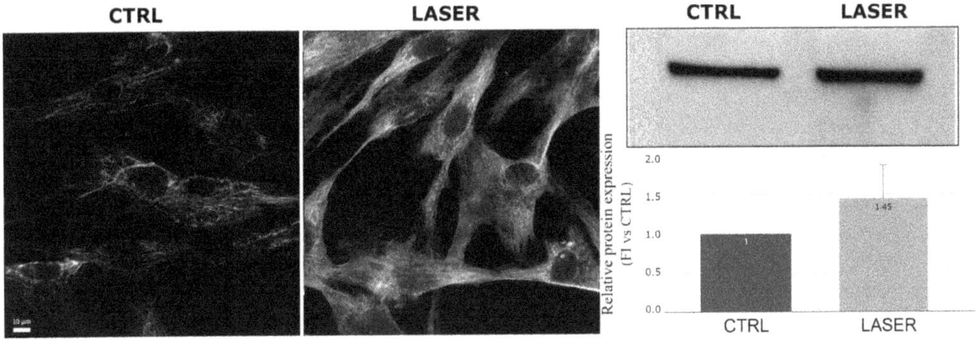

Figure 6. Effects of laser treatment on MMP1 using immunofluorescence and Western blot analysis. Left: control samples; right: laser-treated samples. Western blot bands of control vs. laser-treated samples and graph showing bands' signal intensity ($p > 0.05$).

4. Discussion

In recent years, the field of esthetic medicine has seen significant advancements due in part to the development and application of laser technologies. Among these, red wavelength lasers, typically operating in the range of 600 to 750 nm, and NIR lasers, operating in the range of 780 to 1000 nm, have garnered particular attention for their unique properties and versatile applications [15–20]. The absorption characteristics of red light by biological tissues, along with its ability to penetrate deeper layers of the skin, make these lasers highly effective for a variety of esthetic treatments. Researchers are trying to understand the intricate ways in which these technologies can improve skin health and appearance by delving deeper into the cellular and molecular effects of these non-ablative laser treatments. This approach not only targets esthetic improvements but also contributes to overall tissue health. In our study, we aimed to investigate the effect induced by treatment with an IR laser source (RedTouch laser) on the organization and expression of various cytoskeletal and extracellular matrix proteins in dermal fibroblasts. These cells play a crucial role in tissue repair processes and are responsible for the deposition of a new ECM in response to damage.

Vimentin and actin are essential components of the cytoskeleton, playing critical roles in maintaining cell structure, integrity, and function [22,23]. Our results concerning vimentin expression and organization may indicate that laser treatment affects its spatial organization rather than increasing the total amount of the protein. This suggests that laser radiation might induce changes in the arrangement of vimentin within the cell, which could influence various cellular functions and structural stability. Similarly, since actin has a crucial role in cell movement, division, and intracellular transport, changes in actin polymerization and the rearrangement of the actin filament network could reflect alterations in cytoskeletal dynamics [26]. These changes are potentially related to increased cell mobility and activity induced by laser treatment. Thus, laser treatment might promote a more dynamic and responsive cytoskeletal structure, facilitating enhanced cellular processes such as tissue repair and cell migration. α-SMA is an actin isoform present in myofibroblasts and involved in tissue remodeling [27]. After laser treatment, α-SMA did not show statistically significant variations in either immunofluorescence or the Western blot analysis. This suggests that laser treatment does not significantly alter α-SMA expression, indicating that there is no relevant change in cell contraction and fibroblast-to-myofibroblast transformation, thus reducing the risk of fibrosis during the tissue repair process.

Parallel to the study of cytoskeletal proteins, focal adhesions have also been investigated. Focal adhesions are sub-cellular structures that mediate the regulatory effects of a cell in response to ECM adhesion [28]. The main components of focal adhesions are integrins, which are membrane glycoproteins that play a crucial role in linking the cell to the ECM and in transducing signals from the ECM to the cell. Observations of immunofluorescence images reveal no significant changes in the arrangement or expression of integrins. However, interesting results have been obtained through the study of proteins that interact with integrins, such as paxillin, vinculin, and fibronectin. Paxillin is a signal transduction adaptor protein that localizes to the intracellular surface of focal adhesions and binds directly to integrins. Paxillin plays a central role in coordinating the spatial and temporal actions of the Rho family of small GTPases, which regulate the actin cytoskeleton by recruiting an array of GTPase activators, suppressors, and effector proteins to cell adhesions [29]. Paxillin has been described as an important player in skin fibroblast morphology, and it has been shown that its levels decrease notably during the aging process, consequently leading to reductions in collagen I production and fibroblast contractility [30]. Paxillin also has a crucial role in the system that transduces the signals deriving from the cell–matrix interaction. It is well known that fibroblasts regulate matrix turnover by producing metalloproteinases and ECM components, but the ECM regulates fibroblast morpho-functional phenotypes through biochemical and mechanical cues. An altered cell–matrix interaction due, at least in part, to a decrease in paxillin expression has been associated with aging, and it has been suggested that restoring the paxillin-mediated cell–matrix interaction might play a significant role in anti-aging actions [31]. Our analysis shows that paxillin levels increased significantly after the laser treatment, suggesting that laser radiation could help reverse the decrease in paxillin levels during the aging process.

Vinculin consists of a head domain and a tail domain connected by a flexible linker, enabling it to switch between open and closed conformations. The head domain of vinculin shares several key binding partners and functions with paxillin, with paxillin itself being one of these partners. However, unlike paxillin, which does not directly interact with actin, the tail domain of vinculin can bind directly to actin filaments and interact with actin-binding proteins such as α-actinin [32]. In the literature, a reduction in vinculin has been associated with increased cytoskeletal plasticity and cell motility [33]; this may facilitate cellular remodeling processes and more dynamic responses. The reduction in vinculin observed in our analyses could indicate a change in the dynamics of focal adhesions and in the interaction between the cytoskeleton and the ECM.

Fibronectin is an essential glycoprotein for cell–matrix adhesions as it can bind both integrins on the external side of the plasma membrane and ECM molecules, such as collagen. Recent studies in the literature show that rejuvenated fibroblasts exhibit increased matrix protein deposition, including fibronectin [34]. In agreement with the current literature, our results show a significant increase in fibronectin expression after laser treatment. It is well established that fibronectin, together with other proteins such as heparan sulfate, acts as a scaffold to organize enzymes and substrates for procollagen processing [35]; therefore, we decided to investigate collagen expression.

While the analysis of collagen I expression showed a slight reduction after the laser treatment, a significant increase in collagen III expression was observed. These data should not be seen as contradictory; they can be explained by the fact that the production of collagen I and III occurs at different stages of tissue repair. Initially, collagen III synthesis predominates, which is later replaced by collagen I [36]. As shown in our experiments, the early increase in collagen III induced by the non-ablative laser treatment, along with the slight reduction in collagen I, suggest potential reorganization and increased deposition of the ECM, which may enhance the structure and resilience of skin tissue, therefore confirming its anti-aging effect in the esthetic field. The results confirm the outcomes of a preliminary study [12] and demonstrate the anti-aging effect of the laser source and treatment protocol applied in this study.

Finally, the expression of MMP-1, also known as interstitial collagenase, was evaluated because its main targets are collagen types I, II, and III [37]. Although not reaching statistical significance, the trend toward increased MMP1 expression further supports the hypothesis that laser treatment promotes the reorganization of the matrix and remodeling of collagen fibers.

We conducted a cellular investigation to understand the molecular mechanism involved in the skin rejuvenation process's signaling pathways in accordance with clinical evidence that has already been tested. In fact, in previous investigations, we showed that the RedTouch laser's capacity to interact with collagen fibers makes it a potentially effective therapy for chrono aging [15–17]. In these clinical articles, we compared our non-ablative laser system with other gold-standard treatments for rejuvenation. Ablative lasers vaporize tissue and are therefore more aggressive than non-ablative lasers. Compared to ablative lasers (CO_2/erbium lasers), non-ablative lasers have much fewer possible harmful consequences, and their main advantage is a significant decrease in postoperative downtime.

4.1. Potential Future Applications

The RedTouch device emits a laser in the visible spectrum, and a fractional scanning system allows micro thermal zones (with an average depth of 300/500 microns) to be generated; this thermal column conducts heat to the surrounding areas, causing immediate collagen shrinkage and denaturation with subsequent new collagen formation. Laser-generated thermal columns could positively impact wound healing through antibacterial and anti-inflammatory properties (in addition to neocollagenesis), promoting the release of cytokines and growth factors that are crucial for wound management [37–39].

Injectable hyaluronic acid fillers and laser/light procedures have become increasingly popular for non-invasive facial rejuvenation in many cosmetic practices. As a result, in a preclinical internal laboratory test, we have already examined the possible interaction between the study device and some hyaluronic acid-based fillers using a Raman spectroscopy analysis, showing that the tested device is safe when interacting with hyaluronic acid without changing its molecular structure. From a clinical approach, a 675 nm laser interacts with hyaluronic acid-based fillers to improve its effectiveness in skin rejuvenation procedures (data are in the process of publication). Natural polysaccharide hyaluronic acid, which is generated by fibroblasts during the wound repair stages of proliferation, can mediate cellular signaling, encourage cell migration, and induce morphogenesis and the matrix structure, all of which improve the soft tissue wound healing process [40].

A hydrogel is an innovative medical device that promotes healing due to the presence of hyaluronic acid, protecting the wound by promoting skin re-epithelialization.

Since existing studies indicate that NIR irradiation enhances hydrogels' in vivo antibacterial properties [41–43], we may consider applying our technology to the field of wound healing with a completely new approach; this is one of our next research goals.

4.2. The Limitations of This Study

A limit of this study is that it considered a single treatment, the effects of which were analyzed after 24 h, i.e., at a single time point. Based on the results, further investigations considering repeated treatments and several post-treatment timepoints will be implemented in future research.

5. Conclusions

In conclusion, the results from immunofluorescence and the Western blot analyses indicate that the laser treatment evaluated in this study induces significant changes in human dermal fibroblasts, particularly affecting cytoskeletal organization and the production and reorganization of extracellular matrix (ECM) molecules. The cellular response to this treatment is closely associated with paxillin-mediated signaling pathways, with paxillin acting as a key adaptor protein in focal adhesions and playing an essential role in cell–matrix interactions. The significant increase in paxillin levels observed after laser exposure suggests that this treatment may counteract the decline in paxillin levels associated with aging, potentially benefiting fibroblast morphology, contractility, and ultimately skin structure. This effect may influence matrix turnover and tissue remodeling, thus providing an anti-aging action further corroborated by increases in fibronectin and collagen III, which are both involved in the early stages of tissue repair.

Overall, these findings highlight the potential of laser treatment as a safe and less invasive clinical approach to skin rejuvenation, with meaningful benefits for patient well-being and quality of life. However, further in-depth studies will be necessary to explore the molecular mechanisms underlying these effects and to validate the anti-aging efficacy of this treatment.

Supplementary Materials: The following supporting information can be downloaded at: https://www.mdpi.com/article/10.3390/biomedicines12122713/s1, Figure S1. Laser emission at 675 nm: molecular counteraction of the aging process; Effects of laser treatment on Vimentin using Western blot analysis. Western blot bands of control vs. laser-treated samples and bands' signal intensity graph ($p > 0.05$). Figure S2. Laser emission at 675 nm: molecular counteraction of the aging process; Effects of laser treatment on Alpha-SMA using Western blot analysis. Western blot bands of control vs. laser-treated samples and bands' signal intensity graph ($p > 0.05$). Figure S3. Laser emission at 675 nm: molecular counteraction of the aging process; Effects of laser treatment on Collagen I using immunofluorescence and Fluorescence Intensity analysis. On the left: control samples: on the right laser-treated samples. Fluorescence Signal Intensity graph of control vs. laser-treated samples ($p < 0.05$).

Author Contributions: Conceptualization, L.N., T.Z. and M.M.; methodology, L.N., L.P., F.C. and C.R.; software, L.N., F.C., C.R., S.B., T.Z. and M.M.; validation, L.N., L.P., F.C., I.F., C.R., F.M., S.B., T.Z. and M.M.; formal analysis, L.N., F.C., I.F., C.R., S.B. and M.M.; investigation, L.N., L.P., F.C., C.R., S.B., T.Z. and M.M.; resources, L.N., F.C., C.R., S.B. and M.M.; data curation, L.N., L.P., F.C., I.F., C.R., F.M., S.B., T.Z. and M.M.; writing—original draft preparation, L.N., L.P., I.F. and M.M.; writing—review and editing, L.N., L.P., F.C., I.F., C.R., F.M., S.B., T.Z. and M.M.; visualization, L.N., L.P., F.C., I.F., C.R., F.M., S.B., T.Z. and M.M.; supervision, M.M. and T.Z.; project administration, M.M. and T.Z.; funding acquisition, M.M. and T.Z.; L.N. and L.P. contributed equally to this work. All authors have read and agreed to the published version of the manuscript.

Funding: This research received no external funding.

Institutional Review Board Statement: Not applicable.

Informed Consent Statement: Not applicable.

Data Availability Statement: The data that support the study's findings are available upon reasonable request from the corresponding author.

Conflicts of Interest: Authors T.Z., L.P., F.M., and I.F. were employed by El.En. Group. The remaining authors declare that this research was conducted in the absence of any commercial or financial relationships that could be construed as potential conflicts of interest.

References

1. Hamblin, M.R. Mechanisms and mitochondrial redox signaling in photo-biomodulation. *Photochem. Photobiol.* **2018**, *94*, 199–212. [CrossRef] [PubMed]
2. Chertkova, R.V.; Brazhe, N.A.; Bryantseva, T.V.; Nekrasov, A.N.; Dolgikh, D.A.; Yusipovich, A.I.; Sosnovtseva, O.; Maksimov, G.V.; Rubin, A.B.; Kirpichnikov, M.P. New insight into the mechanism of mitochondrial cytochrome c function. *PLoS ONE* **2017**, *31*, e0178280. [CrossRef]
3. Barolet, D.; Roberge, C.J.; Auger, F.A.; Boucher, A.; Germain, L. Regulation of skin collagen metabolism in vitro using a pulsed 660 nm LED light source: Clinical correlation with a single-blinded study. *J. Invest. Dermatol.* **2009**, *129*, 2751–2759. [CrossRef] [PubMed]
4. Shapiro, M.G.; Homma, K.; Villarreal, S.; Richter, C.P.; Bezanilla, F. Infrared light excites cells by changing their electrical capacitance. *Nat. Commun.* **2012**, *3*, 736. [CrossRef] [PubMed]

5. Passarella, S.; Karu, T. Absorption of monochromatic and narrow band radiation in the visible and near IR by both mitochondrial and non-mitochondrial photo-acceptors results in Photobiomodulation. *J. Photochem. Photobiol.* **2014**, *140*, 344–358. [CrossRef] [PubMed]
6. Farage, M.A.; Miller, K.W.; Elsner, P.; Maibach, H.I. Intrinsic and extrinsic factors in skin ageing: A review. *Int. J. Cosmet. Sci.* **2008**, *30*, 87–95. [CrossRef]
7. Durai, P.C.; Thappa, D.M.; Kumari, R.; Malathi, M. Aging in elderly: Chronological versus photoaging. *Indian J. Dermatol.* **2012**, *57*, 343–352. [CrossRef]
8. Kim, J.M.; Kim, S.Y.; Noh, E.M.; Song, H.K.; Lee, G.S.; Kwon, K.B.; Lee, Y.R. Reversine inhibits MMP-1 and MMP-3 expressions by suppressing of ROS/MAPK/AP-1 activation in UV-stimulated human keratinocytes and dermal fibroblasts. *Exp. Dermatol.* **2018**, *27*, 298–301. [CrossRef]
9. de Moura, J.P.; de Moura Fernandes, É.P.; Lustoza Rodrigues, T.C.M.; Messias Monteiro, A.F.; de Sousa, N.F.; Dos Santos, A.M.F.; Scotti, M.T.; Scotti, L. Targets Involved in Skin Aging and Photoaging and their Possible Inhibitors: A Mini-review. *Curr. Drug Targets* **2023**, *24*, 797–815. [CrossRef]
10. Han, H.S.; Hong, J.K.; Park, S.J.; Park, B.C.; Park, K.Y. A Randomized, Prospective, Split-Face Pilot Study to Evaluate the Safety and Efficacy of 532-nm and 1,064-nm Picosecond-Domain Neodymium:Yttrium-Aluminum-Garnet Lasers Using a Diffractive Optical Element for Non-Ablative Skin Rejuvenation: Clinical and Histological Evaluation. *Ann. Dermatol.* **2023**, *35*, 23–31. [CrossRef]
11. Magni, G.; Pieri, L.; Fusco, I.; Madeddu, F.; Zingoni, T.; Rossi, F. Laser emission at 675 nm: In vitro study evidence of a promising role in skin rejuvenation. *Regen. Ther.* **2023**, *22*, 176–180. [CrossRef] [PubMed]
12. Munavalli, G.S.; Leight-Dunn, H.M. Pilot Study to Demonstrate Improvement in Skin Tone and Texture by Treatment with a 1064 nm Q-Switched Neodymium-Doped Yttrium Aluminum Garnet Laser. *J. Clin. Med.* **2024**, *13*, 1380. [CrossRef] [PubMed]
13. Ross, E.V.; Zelickson, B.D. Biophysics of nonablative dermal remodeling. *Semin. Cutan. Med. Surg.* **2002**, *21*, 251–265. [CrossRef] [PubMed]
14. Nisticò, S.P.; Tolone, M.; Zingoni, T.; Tamburi, F.; Scali, E.; Bennardo, L.; Cannarozzo, G. A new 675 nm laser device in the treatment of melasma: Results of a prospective observational study. *Photobiomodul. Photomed. Laser Surg.* **2020**, *38*, 560–564. [CrossRef] [PubMed]
15. Bonan, P.; Verdelli, A.; Pieri, L.; Fusco, I.; Linpiyawan, R. Facial rejuvenation: A safe and effective treatment with a fractional non-ablative 675 nm laser in Asian population. *J. Cosmet. Dermatol.* **2021**, *20*, 4070–4072. [CrossRef]
16. Cannarozzo, G.; Bennardo, L.; Zingoni, T.; Pieri, L.; Duca, E.D.; Nisticò, S.P. Histological Skin Changes After Treatment with 675 nm Laser. *Photobiomodul. Photomed. Laser Surg.* **2021**, *39*, 617–621. [CrossRef]
17. Piccolo, D.; Crisman, G.; Dianzani, C.; Zalaudek, I.; Fusco, I.; Conforti, C. New 675 nm Laser Device: The Innovative and Effective Non-Ablative Resurfacing Technique. *Medicina* **2023**, *59*, 1245. [CrossRef]
18. Coricciati, L.; Gabellone, M.; Donne, P.D.; Pennati, B.M.; Zingoni, T. The 675-nm wavelength for treating facial melasma. *Skin Res. Technol.* **2023**, *29*, e13434. [CrossRef]
19. Tolone, M.; Bennardo, L.; Zappia, E.; Scali, E.; Nisticò, S.P. New Insight into Nonablative 675-nm Laser Technology: Current Applications and Future Perspectives. *Dermatol. Clin.* **2024**, *42*, 45–50. [CrossRef]
20. Cole, M.A.; Quan, T.; Voorhees, J.J.; Fisher, G.J. Extracellular matrix regulation of fibroblast function: Redefining our perspective on skin aging. *J. Cell. Commun. Signal.* **2018**, *12*, 35–43. [CrossRef]
21. Wang, J.; Zohar, R.; McCulloch, C.A. Multiple roles of alpha-smooth muscle actin in mechanotransduction. *Exp. Cell. Res.* **2006**, *312*, 205–214. [CrossRef] [PubMed]
22. Sharma, S.; Rai, V.K.; Alieva, I.B.; Shakhov, A.S.; Dayal, A.A.; Churkina, A.S.; Parfenteva, O.I.; Minin, A.A. Unique Role of Vimentin in the Intermediate Filament Proteins Family. *Biochemistry* **2024**, *89*, 726–736. [CrossRef]
23. Narang, R.K.; Markandeywar, T.S. Collagen-based formulations for wound healing: A literature review. *Life Sci.* **2022**, *290*, 120096. [CrossRef]
24. Cabral-Pacheco, G.A.; Garza-Veloz, I.; Castruita-De la Rosa, C.; Ramirez-Acuña, J.M.; Perez-Romero, B.A.; Guerrero-Rodriguez, J.F.; Martinez-Avila, N.; Martinez-Fierro, M.L. The Roles of Matrix Metalloproteinases and Their Inhibitors in Human Diseases. *Int. J. Mol. Sci.* **2020**, *21*, 9739. [CrossRef] [PubMed]
25. Huber, F.; Schnauß, J.; Rönicke, S.; Rauch, P.; Müller, K.; Fütterer, C.; Käs, J. Emergent complexity of the cytoskeleton: From single filaments to tissue. *Adv. Phys.* **2013**, *62*, 1–112. [CrossRef]
26. Sawant, M.; Hinz, B.; Schönborn, K.; Zeinert, I.; Eckes, B.; Krieg, T.; Schuster, R. A story of fibers and stress: Matrix-embedded signals for fibroblast activation in the skin. *Wound Repair Regen.* **2021**, *29*, 515–530. [CrossRef]
27. Chen, C.S.; Alonso, J.L.; Ostuni, E.; Whitesides, G.M.; Ingber, D.E. Cell shape provides global control of focal adhesion assembly. *Biochem. Biophys. Res. Commun.* **2003**, *307*, 355–361. [CrossRef]
28. Deakin, N.O.; Turner, C.E. Paxillin comes of age. *J. Cell. Sci.* **2008**, *121*, 2435–2444. [CrossRef]
29. Zheng, Q.; Chen, S.; Chen, Y.; Lyga, J.; Santhanam, U. Critical role of paxillin in aging of human skin. *J. Invest. Dermatol.* **2012**, *132*, 1290–1293. [CrossRef]
30. Skoczyńska, A.; Budzisz, E.; Podgórna, K.; Rotsztejn, H. Paxillin and its role in the aging process of skin cells. *Postep. Hig. Med. Dosw.* **2016**, *70*, 1087–1094. [CrossRef]

31. Legerstee, K.; Geverts, B.; Slotman, J.A.; Houtsmuller, A.B. Dynamics and distribution of paxillin, vinculin, zyxin and VASP depend on focal adhesion location and orientation. *Sci. Rep.* **2019**, *9*, 10460. [CrossRef] [PubMed]
32. Ezzell, R.M.; Goldmann, W.H.; Wang, N.; Parashurama, N.; Ingber, D.E. Vinculin promotes cell spreading by mechanically coupling integrins to the cytoskeleton. *Exp. Cell. Res.* **1997**, *231*, 14–26. [CrossRef] [PubMed]
33. Roy, B.; Yuan, L.; Lee, Y.; Bharti, A.; Mitra, A.; Shivashankar, G.V. Fibroblast rejuvenation by mechanical reprogramming and redifferentiation. *Proc. Natl. Acad. Sci. USA* **2020**, *117*, 10131–10141. [CrossRef] [PubMed]
34. Saunders, J.T.; Schwarzbauer, J.E. Fibronectin matrix as a scaffold for procollagen proteinase binding and collagen processing. *Mol. Biol. Cell.* **2019**, *30*, 2218–2226. [CrossRef] [PubMed]
35. Witte, M.B.; Barbul, A. General principles of wound healing. *Surg. Clin. N. Am.* **1997**, *77*, 509–528. [CrossRef]
36. Pardo, A.; Selman, M. MMP-1: The elder of the family. *Int. J. Biochem. Cell. Biol.* **2005**, *37*, 283–288. [CrossRef]
37. Jere, S.W.; Houreld, N.N.; Abrahamse, H. Role of the PI3K/AKT (mTOR and GSK3β) signalling pathway and photobiomodulation in diabetic wound healing. *Cytokine Growth Factor Rev.* **2019**, *50*, 52–59. [CrossRef]
38. Oyebode, O.; Houreld, N.N.; Abrahamse, H. Photobiomodulation in diabetic wound healing: A review of red and near-infrared wavelength applications. *Cell. Biochem. Funct.* **2021**, *39*, 596–612. [CrossRef]
39. Fernández-Guarino, M.; Bacci, S.; Pérez González, L.A.; Bermejo-Martínez, M.; Cecilia-Matilla, A.; Hernández-Bule, M.L. The Role of Physical Therapies in Wound Healing and Assisted Scarring. *Int. J. Mol. Sci.* **2023**, *24*, 7487. [CrossRef]
40. Kawano, Y.; Patrulea, V.; Sublet, E.; Borchard, G.; Iyoda, T.; Kageyama, R.; Morita, A.; Seino, S.; Yoshida, H.; Jordan, O.; et al. Wound Healing Promotion by Hyaluronic Acid: Effect of Molecular Weight on Gene Expression and In Vivo Wound Closure. *Pharmaceuticals* **2021**, *14*, 301. [CrossRef]
41. Cheng, S.; Wang, H.; Pan, X.; Zhang, C.; Zhang, K.; Chen, Z.; Dong, W.; Xie, A.; Qi, X. Dendritic Hydrogels with Robust Inherent Antibacterial Properties for Promoting Bacteria-Infected Wound Healing. *ACS Appl. Mater. Interfaces.* **2022**, *14*, 11144–11155. [CrossRef] [PubMed]
42. Qi, X.; Xiang, Y.; Cai, E.; Ge, X.; Chen, X.; Zhang, W.; Li, Z.; Shen, J. Inorganic–organic hybrid nanomaterials for photothermal antibacterial therapy. *Coord. Chem. Rev.* **2023**, *496*, 215426. [CrossRef]
43. Zhu, S.; Zhao, B.; Li, M.; Wang, H.; Zhu, J.; Li, Q.; Gao, H.; Feng, Q.; Cao, X. Microenvironment responsive nanocomposite hydrogel with NIR photothermal therapy, vascularization and anti-inflammation for diabetic infected wound healing. *Bioact. Mater.* **2023**, *26*, 306–320. [CrossRef] [PubMed]

biomedicines

MDPI

Article

Novel BODIPY Dyes with a *Meso*-Benzoxadiazole Substituent: Synthesis, Photophysical Studies, and Cytotoxic Activity Under Normoxic and Hypoxic Conditions

Weronika Porolnik [1,2,3], Natalia Karpinska [2], Marek Murias [1], Jaroslaw Piskorz [2,*] and Malgorzata Kucinska [1,*]

1 Chair and Department of Toxicology, Poznan University of Medical Sciences, 3 Rokietnicka Street, 60-806 Poznan, Poland; w.porolnik@op.pl (W.P.); marek.murias@ump.edu.pl (M.M.)
2 Chair and Department of Inorganic and Analytical Chemistry, Poznan University of Medical Sciences, 3 Rokietnicka Street, 60-806 Poznan, Poland; nataliakarpinska20@gmail.com
3 Doctoral School, Poznan University of Medical Sciences, 70 Bukowska Street, 60-812 Poznan, Poland
* Correspondence: piskorzj@ump.edu.pl (J.P.); kucinska@ump.edu.pl (M.K.)

Abstract: Background/Objectives: Novel boron dipyrromethene derivatives with a heterocyclic, benzoxadiazole substituent were obtained as potential candidates for the photodynamic therapy (PDT) of cancers. Photochemical properties (e.g., singlet oxygen generation quantum yields (Φ_Δ), absorption, and emission spectra) and cytotoxic activity studies in normoxic and hypoxic conditions were performed to verify the potential of novel BODIPYs as photosensitizers for PDT. **Methods:** Obtained dyes were characterized using mass spectrometry and various NMR techniques. The relative method with Rose Bengal as a reference and 1,3-diphenylisobenzofuran as a singlet oxygen quencher was used to determine Φ_Δ values. The in vitro studies were conducted on human ovarian carcinoma (A2780) and human breast adenocarcinoma (MDA-MB-231) cells. **Results:** Photochemical studies showed that the presence of benzoxadiazole moiety only slightly affected the localization of the absorption maxima but resulted in fluorescence quenching compared with *meso*-phenyl-substituted analogs. In addition, brominated and iodinated analogs revealed a high ability to generate singlet oxygen. Anticancer studies showed high light-induced cytotoxicity of BODIPYs containing heavy atoms with very low IC_{50} values in the 3.5–10.3 nM range. Further experiments revealed that both compounds also demonstrated phototoxic activity under hypoxic conditions. The most potent cytotoxic effect in these conditions was observed in the iodinated BODIPY analog with IC_{50} values of about 0.3 and 0.4 μM for A2780 and MDA-MB-231 cells, respectively. **Conclusions:** The results of this study highlighted the advantages and some potential drawbacks of BODIPY compounds with heavy atoms and benzoxadiazole moiety as a useful scaffold in medicinal chemistry for designing new photosensitizers.

Keywords: anticancer research; benzofurazan; BODIPY; halogenation; heavy atom effect; hypoxia; oxadiazole; photochemistry; photodynamic therapy; singlet oxygen

1. Introduction

Boron dipyrromethene derivatives (BODIPYs) and their analogs with the carbon atom at 8 position (*meso*) replaced by nitrogen (aza-BODIPY) are important dyes with numerous applications, including photovoltaics, catalysis, fluorescence imaging, and the preparation of various sensors and optical devices [1,2]. BODIPY dyes have been commonly used as fluorescent markers for various biomedical applications, including imaging, disease diagnosis,

https://doi.org/10.3390/biomedicines13020297

and biological process monitoring. They have also been utilized for anticancer treatment as sensitizers in photodynamic therapy (PDT) and photothermal therapy (PTT) [3]. BODIPY dyes are also connected with other fluorophores, such as xanthenes, cyanines, and coumarins, which are used as fluorophores for various biological applications [4]. In addition, BODIPYs are promising building blocks for various materials for fluorescence labeling and detection, photoacoustic imaging, photodynamic and photothermal therapy, and theranostic agents for integrated diagnosis and treatment [5]. Special attention is paid to using BODIPY dyes in PDT [6,7]. This light-activated therapy proved highly efficient against various cancers, microbial infections, and some non-cancerous diseases [8,9]. The mechanism of photodynamic action usually involves the presence of molecular oxygen in the treated environment to produce reactive oxygen species. For this reason, commonly occurring tumors in oxygen deficient, hypoxic conditions are one of the most critical challenges of PDT [10,11].

BODIPY photosensitizers have been extensively studied for many years, and various modifications have been tested to improve their physicochemical and photochemical properties [12]. Heavy atom substitution is one of the most effective modifications for enhancing singlet oxygen quantum yield. The presence of heavy atoms increases the probability of singlet to triplet intersystem crossing by quenching fluorescence and enhancing spin–orbit coupling [6,13]. In our previous work, we demonstrated the activity of BODIPYs substituted at the *meso* position, with N, N-dimethylaminopropoxyphenyl moiety and with the iodine atoms in positions 2 and 6 of the BODIPY's core. We found that the presence of iodine can indeed increase singlet oxygen generation, from 0.02 for a heavy metal-free BODIPY to 0.97 for iodinated derivatives. Furthermore, inserting two iodine atoms can decrease IC_{50} values from ~3 μM to 48 nM [14].

Boron dipyrromethene dyes are commonly substituted at the *meso* position with phenyl or alkyl groups [3,7,15]. It is known that the substitution at the *meso* position of the BODIPY scaffold can change the electronic density upon excitation and modify the compound's properties [16]. The *meso*-substituted phenyl with an electron donor (e.g., amino-phenyl) and electron-withdrawing substituents (e.g., nitro or carboxyl group) can change the physicochemical and photochemical properties, including fluorescence emission, singlet oxygen generation, lipophilicity, and solubility [16–20]. Heterocyclic substituents at *meso* positions are less common, but recently reported examples have shown interesting properties and possible biomedical applications. Shi et al. synthesized *meso*-benzothiazole BODIPYs as fluorescent rotors for sensing lysosomal viscosity changes, which can be used to monitor the autophagy process and related diseases [21]. A *meso*-indol-substituted BODIPY showed high photocytotoxicity against human lung adenocarcinoma A549 cells and was also used for two-photon fluorescence imaging performed in zebrafish [22]. Several BODIPY dyes with heterocyclic substituents such as thiophene, thiazole, and thienothiophene at the *meso* position have been prepared and subjected to antiparasitic studies in *Trypanosoma brucei* and *Leishmania major* [23]. In addition, BODIPYs with *meso*-benzodioxole substituent revealed promising photodynamic inactivation of the *Staphylococcus aureus* bacteria [24].

Oxadiazoles are five-membered heterocycles containing one oxygen and two nitrogen atoms existing in different isomeric forms. There are many examples of 1,2,4- and 1,3,4-oxadiazoles, while 1,2,5-oxadiazoles, also known as furazans, are less common but have recently gained more attention. In addition, furazans can form fused, bicyclic rings such as benzofurazan. Benzofurazan is a chemical scaffold that can balance the lipophilic character with a high degree of polarization and inductive effect, ensured by the oxadiazole moiety. Additionally, furazan itself exhibits a significant electron-withdrawing characteristic [25]. Various chemical and biological applications of furazan-based compounds, including fluorescent probes for biomolecules, as well as antimicrobial and anticancer agents, have been researched. In addition, benzofurazan moiety is found in the structure of isradipine, a dihydropyridine calcium channel antagonist utilized to treat hypertension [25,26]. The derivatives of nitrobenzoxadiazole (nitrobenzofurazan, NBD) have attracted particular attention due to their fluorescent properties, which can be used to label various important molecules in medicinal chemistry and biology [27].

A few examples of dyes from the boron dipyrromethene class possess oxadiazole moiety. Two BODIPYs with one or two NBD substituents linked via styryl group were synthesized as fluorescence probes for hydrogen sulfide (H_2S) detection. These probes were successfully used for hydrogen sulfide imaging in cultured cells and zebrafish [28]. Another fluorescent probe for H_2S, containing an NBD substituent, was synthesized by Ding et al. The broad studies using aqueous solutions, human cultured cells, and zebrafish showed good characteristics for detecting hydrogen sulfide, including a fast response with intensive emission at 580 nm, high selectivity, low detection limit, and low cytotoxicity [29]. NBD moiety was also introduced to the aza-BODIPY dye to obtain the near-infrared fluorescent probe for discriminative detection of cellular thiols, such as cysteine, homocysteine, and glutathione in living cells [30]. A BODIPY dye containing amino-nitro-benzofurazan moiety was used to simultaneously generate singlet oxygen and nitric oxide. The efficient light-induced production of two independent cytotoxic agents and efficient encapsulation in biodegradable nanoparticles make this molecule a promising candidate for the photodynamic treatment of cancers [31]. It is worth mentioning that to the best of our knowledge, there are no BODIPYs with oxadiazole or benzoxadiazole moieties attached directly at the *meso* position.

The presented studies describe novel BODIPYs with heterocyclic, benzoxadiazole substituents at the *meso* position. Photochemical studies and in vitro cytotoxic activity assays towards two human cancer cell lines under normoxic and hypoxic conditions were performed to verify the potential of novel BODIPYs as photosensitizers for anticancer PDT. The chemical scaffolds used in this study and the effect of modifications involving substitution at the *meso* position and halogenation in the BODIPY structure [32] are presented in Figure 1.

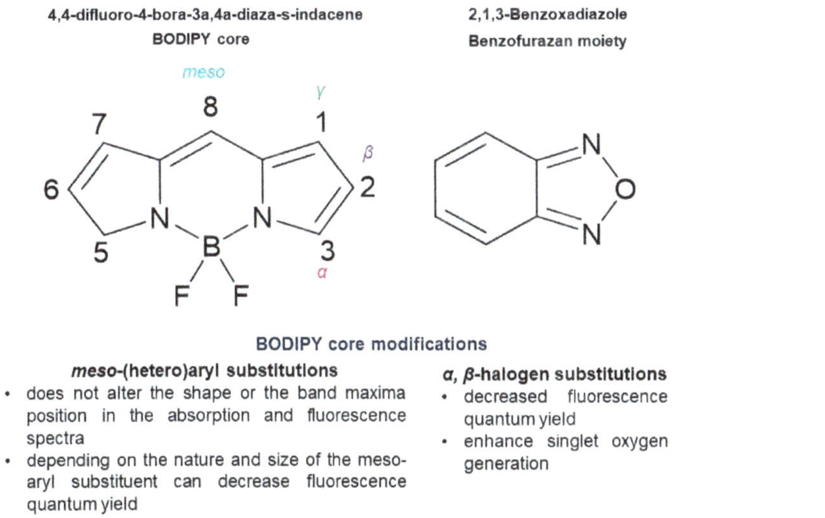

Figure 1. The chemical scaffolds used in this study. The figure also illustrates the possibilities of chemical modification involving the incorporation of aryl-substituent at the *meso* position and halogenation in the BODIPY core.

2. Materials and Methods

2.1. General Procedures

Reactions were conducted under an argon atmosphere on a heat gun-dried glass with Heat-On heating blocks (Radleys, Essex, UK). Aluminum-backed silica gel plates with an F-254 indicator (SiliCycle, Quebec, QC, Canada) were used for thin-layer chromatography (TLC). Solvents were removed at reduced pressure using the R-100 (Buchi Labortechnik, Flawil, Switzerland) and Hei-VAP (Heidolph Instruments, Schwabach, Germany) rotary evaporators. Absorption spectra were acquired with a Shimadzu UV-1900i spectrophotometer (Shimadzu Deutschland, Duisburg, Germany). The AvanceCore 400 spectrometer (Bruker, Billerica, MA, USA) at the Center of Innovative Pharmaceutical Technology at Poznan University of Medical Sciences was utilized for nuclear magnetic resonance spectroscopy (NMR). The ^{1}H spectra were recorded at 400.13 MHz, while ^{13}C at 100.61 MHz. The chemical shift values (δ) were referred to the solvent residual peak ($CDCl_3$) and given as parts per million (ppm). The singlet and doublet of doublet signals were observed and abbreviated as s and dd, respectively. The QTOF Impact HD spectrometer (Bruker, Billerica, MA, USA) at the Center for Advanced Technologies (Adam Mickiewicz University in Poznan, Poland) was used to acquire high-resolution mass spectra (HRMS). The m/z values were presented with the intensity as a percentage of the maximal signal and the difference (Δ ppm) between the identified and calculated mass.

2.2. Synthetic Procedures

2.2.1. Compound **1**, 8-(2,1,3-Benzoxadiazol-4-yl)-1,3,5,7-tetramethyl-4,4′-difluoro-4-bora-3a,4a-diaza-s-indacene

Two drops of trifluoroacetic acid were added to the mixture of 4-formyl-2,1,3-benzoxadiazole (1 mmol, 150 mg) and 2,4-dimethylpyrrole (2 mmol, 0.21 mL) in 50 mL of anhydrous dichloromethane. After stirring at room temperature for 1 h, DDQ (2 mmol, 450 mg) in dichloromethane (25 mL) was added, and the reaction was carried out at room temperature for another hour. Next, N,N-diisopropylethylamine (15 mmol, 2.61 mL) was added, and then boron trifluoride etherate (19 mmol, 2.47 mL) was instilled, and the reaction mixture was stirred for 72 h. After that, purified water (100 mL) was added, and the obtained mixture was filtrated through Celite and extracted. The organic layer was collected, washed with purified water, and dried with anhydrous sodium sulfate. After the evaporation of the solvent, the solid residue was chromatographed in hexane-dichloromethane 1:2 (v/v). The obtained product was crystallized from dichloromethane-n-pentane to produce a red powder of **1** (560 mg, 31%); R_f (dichloromethane: hexane, 1:1 v/v) = 0.15. HRMS (ESI) m/z; found $[M+H]^+$ 367.1527—5% (4.1 ppm); $C_{19}H_{18}BF_2N_4O$ requires $[M+H]^+$ 367.1542; found $[M+Na]^+$ 389.1359—100% (0.5 ppm); $C_{19}H_{17}BF_2N_4ONa$ requires $[M+Na]^+$ 389.1361; found $[M+K]^+$ 405.1090—10% (2.7 ppm); $C_{19}H_{17}BF_2N_4OK$ requires $[M+K]^+$ 405.1101. ^{1}H NMR (400.1 MHz, $CDCl_3$): δ_H, ppm 8.01 (dd, ^3J = 9 Hz, 1 Hz, 1H, ArH), 7.56 (dd, ^3J = 9 Hz, 7 Hz, 1H, ArH), 7.41 (dd, ^3J = 6.5 Hz, 1 Hz, 1H, ArH), 6.00 (s, 2H, H-pyrrole), 2.57 (s, 6H, $-CH_3$), 1.30 (s, 6H, $-CH_3$). ^{13}C NMR (100.6 MHz, $CDCl_3$): δ_C, ppm 14.5; 14.7; 117.6; 121.9; 124.7; 131.2; 131.3; 131.3; 131.6; 133.1; 141.9; 148.8; 149.0; 157.0.

2.2.2. Compound **2a**, 8-(2,1,3-Benzoxadiazol-4-yl)- 2,6-dibromo-1,3,5,7-tetramethyl-4,4′-difluoro-4-bora-3a,4a-diaza-s-indacene

BODIPY **1** (146.5 mg, 0.4 mmol) was added to 30 mL of dichloromethane. Next, N-bromosuccinimide (213.6 mg, 1.2 mmol) dissolved in dichloromethane (10 mL) was instilled into the reaction mixture. After 4 h, the solvent was evaporated, and the solid residue was chromatographed, with dichloromethane: hexane 1:5 (v/v) as the eluent. Precipitation from a dichloromethane solution with n-pentane gave a dark powder of **2a** (145 mg,

69%). R_f (dichloromethane: hexane, 1:1 *v/v*) = 0.53. HRMS (ESI) *m/z*; found $[M\text{-}F]^+$ 502.9674—0.5% (3.2 ppm); $C_{19}H_{15}BBr_2FN_4O$ requires $[M\text{-}F]^+$ 502.9690; found $[M+H]^+$ 522.9729—0.3% (4.4 ppm); $C_{19}H_{16}BBr_2F_2N_4O$ requires $[M+H]^+$ 522.9752; found $[M+Na]^+$ 544.9562—50% (1.7 ppm); $C_{19}H_{15}BBr_2F_2N_4ONa$ requires $[M+Na]^+$ 544.9571; found $[M+K]^+$ 560.9295—4.3% (2.9 ppm); $C_{19}H_{15}Br_2BF_2N_4OK$ requires $[M+K]^+$ 560.9311. ^{1}H NMR (400.1 MHz, $CDCl_3$): δ_H, ppm 8.07 (dd, ^3J = 9 Hz, 1 Hz, 1H, ArH), 7.59 (dd, ^3J = 9 Hz, 7 Hz, 1H, ArH), 7.41 (dd, ^3J = 7 Hz, 1 Hz, 1H, ArH), 2.63 (s, 6H, $\text{-}CH_3$), 1.30 (s, 6H, $\text{-}CH_3$). ^{13}C NMR (100.6 MHz, $CDCl_3$): δ_C, ppm 13.9; 14.0; 112.7; 118.4; 124.2; 130.3; 131.4; 131.9; 133.6; 139.5; 148.7; 149.1; 155.7.

2.2.3. Compound **2b**, 8-(2,1,3-Benzoxadiazol-4-yl)- 2,6-diiodo-1,3,5,7-tetramethyl-4,4′-difluoro-4-bora-3a,4a-diaza-s-indacene

BODIPY **1** (146 mg; 0.4 mmol) was dissolved in anhydrous ethanol (10 mL), and iodine (152 mg; 0.6 mmol) was added. Next, the iodic acid solution (141 mg; 0.8 mmol) in water (1 mL) was instilled, and the reaction mixture was stirred at 70 °C. The consumption of the substrate was evaluated using TLC, and after 45 min, the solvent was evaporated. The solid residue was chromatographed using silica gel, with dichloromethane: hexane 1:5 to 2:1 *v/v* as the eluent. The obtained product was crystallized from dichloromethane-*n*-pentane to give dark crystals of **2b** (196 mg; 79%) R_f (dichloromethane: hexane, 1:1 *v/v*) = 0.38. HRMS (ESI) *m/z*; found $[M-F]^+$ 598.9404—0.8% (1.3 ppm); $C_{19}H_{15}BFBr_2N_4O$ requires $[M-F]^+$ 598.9412; found $[M+H]^+$ 618.9469—0.7% (0.8 ppm); $C_{19}H_{16}BBr_2F_2N_4O$ requires $[M+H]^+$ 618.9474; found $[M+Na]^+$ 640.9296—62% (0.3 ppm); $C_{19}H_{15}BF_2I_2N_4ONa$ requires $[M+Na]^+$ 640.9294; found $[M+K]^+$ 656.9039—100% (0.8 ppm); $C_{19}H_{15}BF_2I_2N_4OK$ requires $[M+K]^+$ 656.9034. ^{1}H NMR (400.1 MHz, $CDCl_3$): δ_H, ppm 8.07 (dd, ^3J = 9 Hz, 1 Hz, 1H, ArH), 7.59 (dd, ^3J = 9 Hz, 7 Hz, 1H, ArH), 7.40 (dd, ^3J = 7 Hz, 1 Hz, 1H, ArH), 2.67 (s, 6H, $\text{-}CH_3$), 1.32 (s, 6H, $\text{-}CH_3$). ^{13}C NMR (100.6 MHz, $CDCl_3$): δ_C, ppm 16.4; 17.2; 86.7; 118.3; 124.6; 131.2; 131.4; 131.8; 132.8; 144.2; 148.7; 149.1; 158.4.

2.3. Absorption and Emission Properties

The UV-Vis spectra of BODIPYs **1**, **2a**, and **2b** were recorded on a Shimadzu UV-1900i spectrophotometer (Shimadzu Deutschland, Duisburg, Germany), while emission spectra were recorded using a Jasco 6200 spectrofluorimeter (Jasco, Tokyo, Japan). The absorption measurements were performed in triplicate, with molar absorption coefficients given as mean values ± SD. The fluorescence quantum yield measurements were conducted in methanol solutions using a fluorescein solution in 0.1 M NaOH (Φ_{Fref} = 0.92 [33]) as a reference for BODIPY **1** and the Rhodamine B solution in ethanol (Φ_{Fref} = 0.97 [34]) in the cases of BODIPYs **2a** and **2b** [35–37]. The diluted solutions of BODIPYs and references were excited at 500 nm for BODIPY **1** and 530 nm for **2a** and **2b**. The fluorescence quantum yield (Φ_F) values were calculated following previously elaborated procedures [17,38] from three experiments and presented as mean ± SD.

2.4. Photosensitized Production of Singlet Oxygen

The singlet oxygen generation quantum yields (Φ_Δ) were established using the relative chemical trapping method [17,35,39], with Rose Bengal as a reference ($\Phi_{\Delta MeOH}$ = 0.76 [40]) and 1,3-diphenylisobenzofuran (DPBF, Aldrich) as a singlet oxygen quencher. The concentration of DPBF was set to about 3×10^{-5} M, and its decomposition was monitored as absorbance decreased at 409 nm. The methanol solutions of BODIPYs **1**, **2a**, and **2b** or Rose Bengal and DPBF in 1 cm quartz cells were exposed to light at 520 nm from the xenon arc lamp and a M250/1200/U monochromator (Optel, Opole, Poland). The light intensity measured using a PM100D meter with an S120VC sensor (Thorlabs GmbH, Bergkirchen, Germany) equaled 0.5 mW/cm^2. The UV-Vis spectra were recorded at the specific time points of irradiation using a Shimadzu UV-1900i spectrophotometer to determine the rate of DPBF oxidation upon interaction with singlet oxygen generated from tested BODIPYs and Rose Bengal. The Φ_Δ values were calculated following previously elaborated procedures [17,38] and presented as mean values from three measurements ± SD.

2.5. Cytotoxic Activity

2.5.1. Cell Culture

The reagents used for in vitro studies were obtained from Sigma Aldrich (St. Louis, MO, USA) and included Dulbecco's modified Eagle's medium (DMEM), fetal bovine serum (FBS), a penicillin–streptomycin–L-glutamine solution, Dulbecco's phosphate-buffered saline (DPBS), trypsin–EDTA, molecular-grade dimethyl sulfoxide (DMSO), and 3-(4, 5-dimethylthiazol-2-yl)-2,5-diphenyltetrazolium bromide (MTT).

The DMSO used as a solvent for formazan crystals was purchased from Avantor Performance Materials (Gliwice, Poland).

The human ovarian carcinoma cell line (A2780) and human breast adenocarcinoma (MDA-MB-231) were purchased from the European Collection of Authenticated Cell Cultures (ECACC, Salisbury, UK) and American Type Culture Collection (ATCC, Manassas, VA, USA), respectively. Both cell lines were cultured in DMEM supplemented with 10% (*v*/*v*) FBS, 2 mM L-glutamine, 1% (*v*/*v*) penicillin (10,000 U), and a streptomycin solution (10 mg/mL) in a humidified atmosphere of 5% CO_2 at 37 °C.

The tested compounds were dissolved in DMSO at a concentration of 10 mM for the preliminary study. In contrast, to determine the IC_{50} values, the compounds were dissolved at a concentration of 1 mM. Stock solutions were stored in the dark at −20 °C. The cytotoxicity of each photosensitizer was evaluated under both dark and light conditions.

Hypoxia experiments were conducted using the hypoxia station (Whitley H35 Hypoxystation; Don Whitley Scientific, Bingley, UK). The hypoxia station ensured stable and monitored experimental conditions by maintaining oxygen at 1%, carbon dioxide at 5%, nitrogen at 94%, a humidified atmosphere, and a temperature of 37 °C. All reagents used for hypoxia experiments were incubated under hypoxic conditions before being added to the cells.

2.5.2. Cell Viability Under Normoxic and Hypoxic Conditions

In the preliminary experiments conducted under normoxic and hypoxic conditions, the cytotoxic potential of the tested compounds was evaluated on MDA-MB-231 cells (seeded at a density of 15×10^3 cells per well)at concentrations of 0.1 μM, 1 μM, and 10 μM. As a negative control, DMSO was used at a concentration of 0.1% (*v*/*v*) in a cell culture medium. Cells were treated with compounds for 24 h. Following irradiation, the cells were washed twice with DPBS and irradiated at a wavelength of 525 nm with InGaN-based High Power LED Multi-Chip Emitters (RoithnerLaserTechnik, Vienna, Austria). The PM16-130 power meter with a slim photodiode sensor (Thorlabs, Newton, NJ, USA) was used to measure light power before each experiment, and the light dose was 2 J/cm^2. Each experiment included the following groups: cells not treated with PS and not exposed to light (untreated control), cells exposed only to light, cells exposed to the PS without irradiation (to determine dark toxicity), and cells treated with PS and exposed to light (to determine light toxicity). The cells were incubated for 24 h after irradiation at standard cell culture conditions, and the MTT assay was performed according to the previously published protocol [14,41]. The absorbance was measured at a wavelength of 570 nm using a plate reader (Biotek Instruments, Elx-800, Winooski, VT, USA).

For hypoxic conditions, the cells were transferred to the hypoxia station 24 h before adding the compounds. All experimental procedures, including treatment and irradiation, were performed under hypoxic conditions.

To determine the IC_{50} values, the A2780 and MDA-MB-231 cells were seeded at a density of 15×10^3 cells on 96-well plates and incubated overnight at standard cell culture conditions. Cells were treated with compound **1** at concentrations of 100 nM, 200 nM, 500 nM, 600 nM, 800 nM, and 1000 nM and compounds **2a** and **2b** at a concentration range of 0.5 nM, 1 nM, 5 nM, 10 nM, 20 nM, and 50 nM for 24 h. Similar to preliminary experiments, DMSO was used as a control, and the concentration in the cell culture medium did not exceed 0.1%. For hypoxic conditions, the tested concentration range was 100 nM, 200 nM, 500 nM, 600 nM, 800 nM, and 1000 nM for all compounds. The irradiation and MTT assays were performed as a preliminary study.

The IC_{50} values were calculated using the GraphPad Prism 8 software (GraphPad Software, Inc., La Jolla, CA, USA). The variable slope (four parameters) analysis was used to calculate dose and response curves and determine IC_{50} values. To determine IC_{50} values, three biological replications were performed for each compound.

2.6. Statistical Analysis

Statistical analysis was performed using GraphPad Prism® 8 (GraphPad Software, Inc., La Jolla, CA, USA). One-way ANOVA followed by Dunnett's post hoc test was used to determine significance; $p < 0.05$ was considered significant.

3. Results and Discussion

3.1. Synthesis and Characterization

Novel BODIPY **1** possessing a *meso*-benzofurazan substituent was synthesized from 2,4-dimethylpyrrole and 4-formyl-2,1,3-benzoxadiazole in a three-step reaction using trifluoroacetic acid, 2,3-dichloro-5,6-dicyano-1,4-benzoquinone (DDQ), N,N-diisopropylethy lamine (DIPEA), and boron trifluoride etherate in dichloromethane following similar procedures for the synthesis of phthalimide-substituted BODIPYs [42,43]. In the next steps, BODIPY **1** was brominated in positions 2 and 6 of the BODIPY core using N-bromosuccinimide (NBS) to give derivative **2a** and then iodinated with the mixture of I_2 and HIO_3, thereby resulting in the synthesis of compound **2b**, similar to the previously described procedure [44] (Scheme 1).

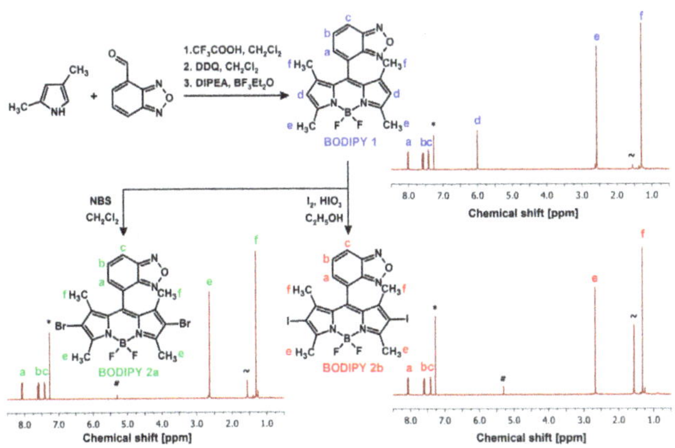

Scheme 1. Synthesis of BODIPYs **1**, **2a**, and **2b** with their ^1H NMR spectra in $CDCl_3$. The letters a-f assign NMR signals to the hydrogen atoms in the BODIPY structures. The residual peaks of chloroform, dichloromethane, and water are indicated by symbols *, #, and ~, respectively.

Novel compounds **1**, **2a**, and **2b** were characterized using high-resolution mass spectrometry (HRMS), UV-Vis spectrophotometry, and various NMR techniques. The ^{1}H NMR spectrum of BODIPY **1** showed six signals (Scheme 1). The signals at 7.41, 7.56, and 8.01 ppm are derived from the phenyl group of a benzoxadiazole substituent at the BODIPY *meso* position. The singlets at 1.30 and 2.57 ppm belong to protons of methyl groups at positions 1 and 7 of the BODIPY core. Another singlet at 6.00 can be assigned to the hydrogen atoms at positions 2 and 6 of the BODIPY core. The NMR spectra of BODIPYs **2a** and **2b** differ from the spectrum of BODIPY **1** due to the lack of the signal at 6.00 ppm, which confirms the introduction of bromine and iodine atoms at positions 2 and 6 of the BODIPY core (Scheme 1). A detailed analysis of the NMR spectra can be found in Supplementary Information (Figures S1–S9, Tables S1–S3).

3.2. Absorption and Emission Properties

The absorption properties of BODIPYs **1**, **2a**, and **2b** were established in various solvents, including dichloromethane, methanol, ethanol, and dimethyl sulfoxide. Figure 2 presents the UV-Vis absorption spectra of the studied dyes in dichloromethane (DCM), while Figure S13 of Supplementary Information includes the spectra in other solvents.

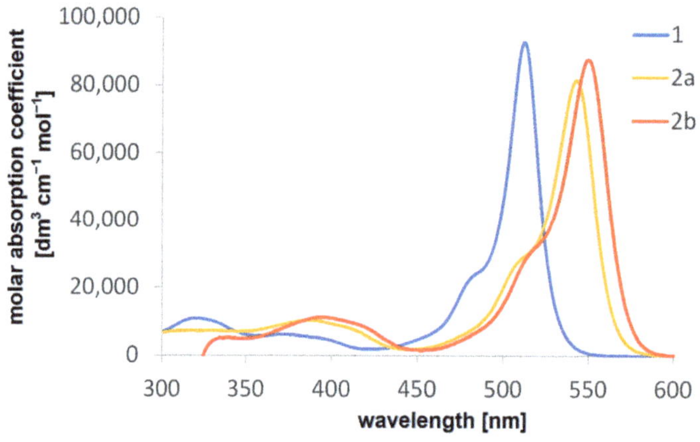

Figure 2. Absorption spectra of compounds **1**, **2a**, and **2b** in dichloromethane.

Table 1 contains the absorption maxima (λ_{Abs}) with logarithms of molar absorption coefficients (log ε) in dichloromethane, whereas Table S4 of Supplementary Information presents data for other solvents. The UV-Vis spectra revealed intensive bands in the 450–600 nm range for halogenated BODIPYs **2a** and **2b** and between 425 and 550 nm for BODIPY **1**. These bands can be assigned to the $S_0 \rightarrow S_1$ transition. The lower intensity bands corresponding to the $S_0 \rightarrow S_2$ transition were found between 300 and 425 nm for BODIPY **1** and 350–450 nm for BODIPYs **2a** and **2b**. Data from the literature suggested that *meso*-aryl substitutions in the BODIPY core did not remarkably alter the absorbance and fluorescence spectra (Figure 1). In this study, we also showed that benzofurazan moiety at the *meso* position slightly affected the localization of the absorption maxima, causing only about a 10 nm shift towards longer wavelengths compared with *meso*-phenyl substituted analogs [14,15]. In addition, the bathochromic shift in absorption bands for compounds **2a** ($\lambda_{Abs\ in\ DCM}$ = 542 nm) and **2b** ($\lambda_{Abs\ in\ DCM}$ = 549 nm), in comparison to BODIPY **1** ($\lambda_{Abs\ in\ DCM}$ = 512 nm), results from the presence of iodine and bromine atoms in the BODIPY core. Notably, all studied compounds revealed high log ε values from 4.67 to 4.99 (Table 1 and Table S4 of Supplementary Information).

Table 1. UV-Vis absorption maxima (λ_{Abs}), logarithms of molar absorption coefficients (log ε), emission wavelengths (λ_{em}), Stokes shifts ($\Delta\lambda$), quantum yields of fluorescence (Φ_F), and singlet oxygen generation yields (Φ_Δ) of BODIPYs **1**, **2a** and **2b** in methanol.

Compound	λ_{Abs} [nm]	log $\varepsilon \pm$ SD	λ_{em} [nm]	$\Delta\lambda$ [nm]	$\Phi_F \pm$ SD	$\Phi_\Delta \pm$ SD
1	512	4.98 ± 0.04	520	8	0.002 ± 0.001	0.02 ± 0.01
2a	542	4.91 ± 0.03	552	10	0.097 ± 0.004	0.55 ± 0.01
2b	549	4.94 ± 0.05	563	14	0.010 ± 0.001	0.63 ± 0.01

The emission spectra of benzofurazan BODIPYs were recorded in methanol (Figure 3), and the obtained fluorescence data, including emission maxima (λ_{em}), Stokes shifts ($\Delta\lambda$), and fluorescence quantum yields (Φ_F), are summarized in Table 1. Regarding the fluorescence quantum yields, all compounds exhibited low Φ_F values, with the lowest value calculated for compound **1**. Bumagina et al. suggested that *meso*-aryl substitutions can decrease fluorescence quantum yields [32]. Our data indicate that a *meso*-aryl substitution with benzofurazan moiety may further reduce fluorescence quantum yields.

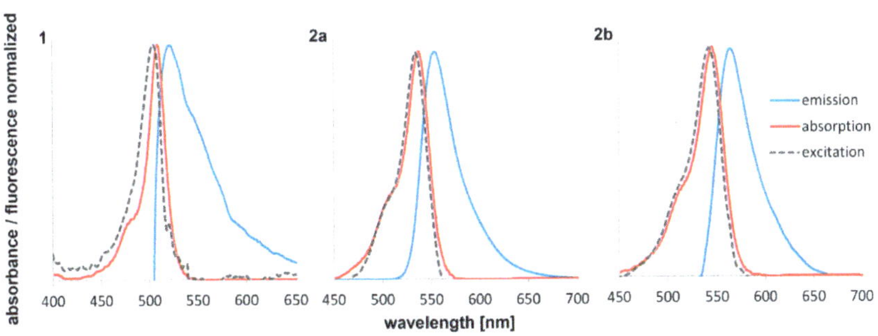

Figure 3. Long-wavelength absorption, excitation, and emission bands for BODIPYs **1**, **2a**, and **2b** in methanol.

The tested BODIPYs revealed a weak fluorescence, with the emission maxima at 520, 552, and 563 nm for BODIPYs **1**, **2a**, and **2b**, respectively. The fluorescence quenching by oxadiazole moiety in a benzofurazan substituent is responsible for the minimal fluorescence observed for unsubstituted derivative **1** [30,45]. In contrast, the *meso*-phenyl-substituted BODIPYs are known for their high fluorescence [14,15,46]. However, the benzofurazan-derived fluorescence quenching effect was reduced by heavy atoms, as BODIPYs **2a** and **2b** revealed a higher fluorescence intensity than the unsubstituted analog **1**. Also, the lower fluorescence intensity of iodinated dye **2b** than brominated **2a** agrees with the heavy atom effect caused by the presence of these atoms in positions 2 and 6 of the BODIPY core [6,44,47,48].

3.3. Singlet Oxygen Generation Measurements

Singlet oxygen quantum yields (Φ_Δ) were determined using a relative method, where Rose Bengal was used as a photosensitizer standard and 1,3-diphenylisobenzofuran (DPBF) as a singlet oxygen quencher. The solutions of compounds **1**, **2a**, **2b**, or Rose Bengal with DPBF were prepared in methanol and irradiated with 520 nm green light. During irradiation, DPBF decomposes and oxidizes, which is observed as decreasing absorption at a wavelength of about 410 nm. The changes in the UV-Vis spectra during measurements are shown in Figure 4, whereas first-order plots of DPBF degradation by singlet oxygen are presented in Figure S14 of Supplementary Information. The calculated values of Φ_Δ are presented in Table 1. Compound **1** showed a low ability to generate singlet oxygen, with the value of Φ_Δ equal to 0.02, compared to the value obtained for the standard substance, Rose Bengal (Φ_Δ = 0.76 [40]). On the other hand, derivatives **2a** and **2b** showed a much higher singlet oxygen generation efficiency with the values of Φ_Δ = 0.55 and 0.63. The significantly higher ability to generate singlet oxygen by the compound containing iodine or bromine atoms in positions 2 and 6 of the BODIPY ring is associated with the heavy atom effect. The presence of iodine or bromine atoms at these positions causes the light-excited molecule to transition from the singlet state to the triplet state, where the molecule can react efficiently with molecular oxygen, forming singlet oxygen with high yields [44,49,50].

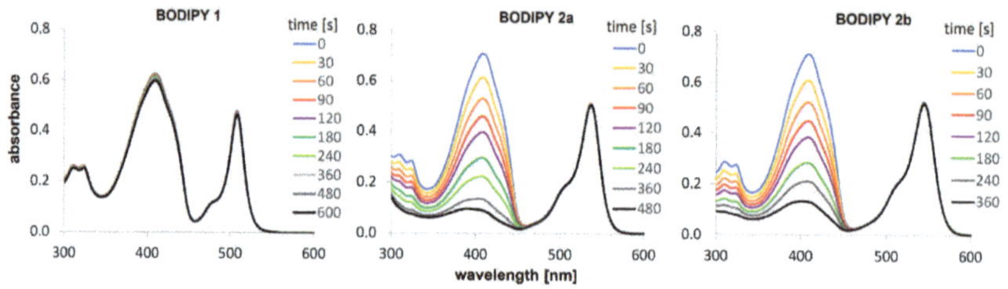

Figure 4. The time-dependent UV-Vis spectra for 1,3-diphenylisobenzofuran (DPBF) upon irradiation at a wavelength of 520 nm with BODIPYs **1**, **2a**, and **2b** in methanol.

3.4. *Cell Viability Evaluation Using the MTT Assay*

Such features of photosensitizers from the BODIPY family, including stability under various environmental conditions, high molar extinction coefficients, and resistance to photobleaching, make these compounds excellent candidates for PDT applications [6]. Moreover, the BODIPY core is a versatile scaffold that can be modified to improve photophysical and photochemical properties, thereby enhancing biological activity [6,32]. This assertion is supported by several in vitro studies in which BODIPYs demonstrated excellent phototoxic activity, with IC_{50} values in the nanomolar range [14,51,52]. As presented in Figures 5 and 6, BODIPYs **1**, **2a**, and **2b** exerted cytotoxic activity towards the A2780 and MDA-MB-231 cell lines after irradiation and did not affect cell viability without light exposure. The results of the preliminary experiment used to determine the concentration range for IC_{50} value determination are presented in Figure S15. The morphology of the A2780 and MDA-MB-231 cells after treatment with BODIPYs **1**, **2a**, and **2b** is shown in Figures S16–S21. The statistical analysis and significance of the obtained results are presented in Supplementary Materials (Figures S22–S25).

The most potent activity was observed for compound **2b**, with IC_{50} values of 3.55 $\pm$ 0.44 nM and 6.47 $\pm$ 1.73 nM for the A2780 and MDA-MB-231 cells, respectively (Table 2).

Thus, our studies confirmed that the presence of iodine atoms at positions 2 and 6 can enhance the cytotoxic activity, probably due to the higher singlet oxygen generation. The insertion of bromine at positions 2 and 6 also ensures prominent cytotoxic activity, with IC_{50} values of 6.08 $\pm$ 0.56 nM and 10.26 $\pm$ 1.06 nM for the A2780 and MDA-MB-231 cells (Table 2). The absence of a heavy atom in compound **1** resulted in a lower activity compared to compounds **2a** and **2b**; however, the IC_{50} values for both cell lines were in the nanomolar concentration range. Badon and co-workers also observed a similar effect for a diaminophenyl-functionalized cationic BODIPY [53]. The authors demonstrated that the iodinated BODIPY exhibited a higher singlet oxygen quantum yield (0.55) than its brominated counterpart (0.13). The authors also found that the iodinated and brominated derivatives showed potent cytotoxic activity against human cervical adenocarcinoma (HeLa) and human breast adenocarcinoma (MCF-7) cells. The iodinated derivative showed

IC_{50} values of 64 nM and 48.57 nM for HeLa and MCF-7 cells, respectively, while the brominated derivative exhibited IC_{50} values of 59.27 nM and 49.35 nM for HeLa and MCF-7 cells, respectively. On the other hand, non-halogenated counterparts did not affect cell viability [53]. In our studies, the non-halogenated derivative **1** decreased the cell viability with IC_{50} values of 182.72 ± 25.81 and 279.34 ± 10.30 nM for the A2780 and MDA-MB-231 cells, respectively, despite the singlet oxygen quantum yield reaching a value of 0.02 (approximately 30 times lower than that of **2b**). Interestingly, there is also approximately a 30-fold difference in the IC_{50} values between compounds **1** and **2b**.

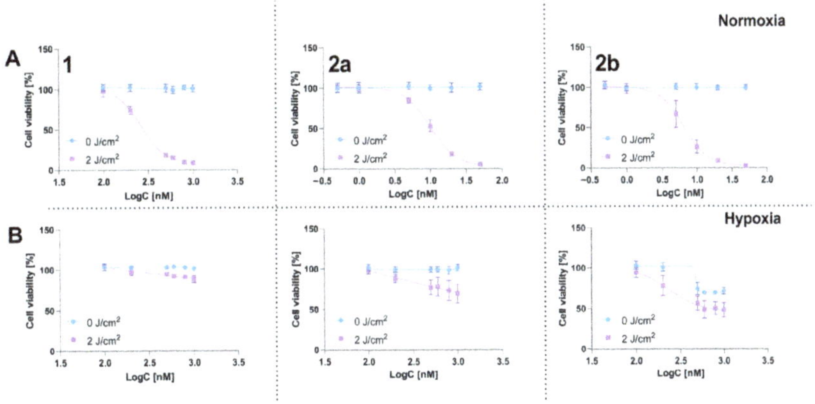

Figure 5. The dose–response curves for compounds **1**, **2a**, and **2b** against MDA-MB-231 cells. Cell viability was measured using the MTT assay 24 h after irradiation (phototoxicity) or without irradiation (dark toxicity) under normoxic (**Panel A**) and hypoxic (**Panel B**) conditions. Data are presented as mean values ± SD, calculated from three independent experiments (except for compound **1** under hypoxic conditions, which was repeated once due to its lack of cytotoxic effect).

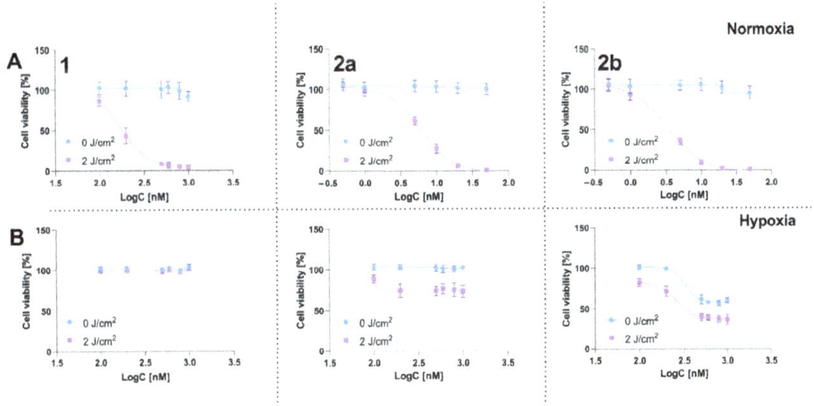

Figure 6. The cytotoxic activity of compounds **1**, **2a**, and **2b** towards A2780 cells. (**Panel A**) presents the results under normoxic conditions, while (**Panel B**) presents the results under hypoxic conditions. Cell viability was measured using the MTT assay 24 h after irradiation (phototoxicity) and without exposure to light (dark toxicity). Data are presented as mean values ± SD, calculated from three independent experiments (except for compound **1** under hypoxic conditions, which was repeated once due to its lack of cytotoxic effect).

Table 2. The IC_{50} values for tested compounds under normoxic conditions against A2780 and MDA-MB-231 cells. Data presented as mean ± SD from three independent experiments.

	IC_{50} [nM]			
	A2780		MDA-MB-231	
	0 J/cm^2	2 J/cm^2	0 J/cm^2	2 J/cm^2
1	>1000	182.72 ± 25.81	>1000	279.34 ± 10.30
2a	>50	6.08 ± 0.56	>50	10.26 ± 1.06
2b	>50	3.55 ± 0.44	>50	6.47 ± 1.73

Since hypoxia is one of the well-known factors that can limit PDT effectiveness, we tested all compounds under hypoxic conditions. Compound **1** was inactive under hypoxic conditions in both cell lines in the concentration range corresponding to the normoxic conditions (Figures 5 and 6). The results obtained under normoxic conditions exhibited the same trends as those under hypoxic conditions. The most potent cytotoxic effect was observed for compound **2b** with IC_{50} values of 293.77 ± 66.76 nM and 429.00 ± 11.72 nM for A2780 and MDA-MB-231 cells (Table 3). However, it should be noted that under hypoxic conditions at the tested concentration range, compound **2b** also affects cell viability without irradiation. At concentrations of 1000 nM, 800 nM, 600 nM, and 500 nM, compound **2b** decreased cell viability to ~60% in A2780 cells (Figure 6, Table S5). Compound **2b** demonstrated a more potent effect without irradiation on ovarian cancer cells under hypoxic conditions since, for MDA-MB-231 cells, this compound decreased cell viability to ~70% compared to the control cells (Figures 5 and 6, Table S5). Compound **2a** under hypoxic conditions reduced cell viability after irradiation at the highest dose to ~70% for both cell lines (Table S6), while it did not affect cell viability without irradiation (Table S5). Therefore, it can be concluded that bromine substitution led to a less photocytoxic effect but also exerted a less cytotoxic effect. Although the generation of singlet oxygen is lower for BODIPYs **2a** and **2b** (0.55 and 0.63, respectively), compared to the already described N,N-dimethylaminopropoxyphenyl derivative in the literature, both compounds also demonstrated activity under hypoxic conditions. In contrast, the BODIPY with the N,N-dimethylaminopropoxyphenyl group at the *meso* position only exerted cytotoxic activity towards the human prostate carcinoma (LNCaP) cell line under normoxic conditions and was inactive in hypoxia [14]. Therefore, considering singlet oxygen generation, the phototoxic activity of compounds **2a** and **2b** could be related to both type I and type II photodynamic reactions in their phototoxic mechanism of action. However, further studies are needed to verify this hypothesis.

Table 3. The IC_{50} values for tested compounds under hypoxic conditions against A2780 and MDA-MB-231 cells. Data presented as mean ± SD from at least two independent experiments.

	IC_{50} [nM]			
	A2780		MDA-MB-231	
	0 J/cm^2	2 J/cm^2	0 J/cm^2	2 J/cm^2
1	>1000	>1000	>1000	>1000
2a	>1000	>1000	>1000	>1000
2b	>1000	293.77 ± 66.76	>1000	429.00 ± 11.72

4. Conclusions

A simple procedure for synthesizing novel boron dipyrromethene derivatives with a heterocyclic, benzoxadiazole substituent at the *meso* position was elaborated. Obtained dyes were characterized using mass spectrometry and various NMR techniques and subjected to photochemical and cytotoxicity studies. The presence of benzoxadiazole moiety only slightly affected the localization of the absorption maxima but resulted in fluorescence quenching compared with *meso*-phenyl-substituted analogs. In addition, brominated and iodinated analogs revealed a high ability to generate singlet oxygen. The in vitro studies using human ovarian carcinoma (A2780) and human breast adenocarcinoma (MDA-MB-231) cells revealed high light-induced cytotoxicity of BODIPYs containing heavy atoms with very low IC_{50} values in the 3.5–10.3 nM range. Further studies revealed that both compounds demonstrated phototoxic activity also under hypoxic conditions. The most potent cytotoxic effect in these conditions was observed for the iodinated BODIPY analog with IC_{50} values of about 0.3 and 0.4 μM for the A2780 and MDA-MB-231 cells, respectively. However, it should be noted that this compound also affects cell viability without irradiation under hypoxic conditions at the tested concentration range. The brominated analog showed less phototoxic effect but also exerted less cytotoxic effect itself. Nevertheless, both BODIPYs with heavy atoms are promising candidates for further research. The results of this study highlighted the advantages and some potential drawbacks of BODIPY compounds with heavy atoms and benzoxadiazole moiety as a useful scaffold in medicinal chemistry for designing new photosensitizers. However, more detailed studies are required to fully characterize the potential of the tested compounds, including mechanistic and functional analyses to elucidate their molecular mechanisms of action. This knowledge could also help identify and overcome potential resistance mechanisms, particularly under hypoxic conditions. Moreover, preclinical in vivo studies could further define the applicability of the tested BODIPYs.

Supplementary Materials: The following supporting information can be downloaded at https://www.mdpi.com/article/10.3390/biomedicines13020297/s1, Figure S1: ^{1}H NMR of BODIPY **1** in deuterated chloroform; Figure S2: ^{13}C NMR of BODIPY **1**; Figure S3: ^{1}H and (^{13}C) chemical shift values [ppm] and key correlations observed in NMR spectra of BODIPY **1**; Figure S4: ^{1}H NMR of BODIPY **2a** in deuterated chloroform; Figure S5: ^{13}C NMR of BODIPY **2a**; Figure S6: ^{1}H and (^{13}C) chemical shift values [ppm] and key correlations observed in NMR spectra of **2b**; Figure S7: ^{1}H NMR of BODIPY **2b** in deuterated chloroform; Figure S8: ^{13}C NMR of BODIPY **2b**; Figure S9: ^{1}H and (^{13}C) chemical shift values [ppm] and key correlations observed in NMR spectra of **2b**; Figure S10: HRMS of BODIPY **1**; Figure S11: HRMS of BODIPY **2a**; Figure S12: HRMS of BODIPY **2b**; Figure S13: Absorption spectra of compounds **1**, **2a**, and **2b**; Figure S14: The first-order plots for the oxidation of the DPBF for BODIPYs **1**, **2a**, and **2b** in methanol; Figure S15: The preliminary results for compounds **1**, **2a**, and **2b** performed in normoxic conditions towards MDA-MB-231 cells; Figure S16: The A2780 morphology after treatment with BODIPY **1**; Figure S17: The A2780 cells morphology after treatment with BODIPY **2a**; Figure S18: The A2780 morphology after treatment with BODIPY **2b**; Figure S19: The MDA-MB-231 cells morphology after treatment with BODIPY 1; Figure S20: The MDA-MB-231 cells morphology after treatment with BODIPY **2a**; Figure S21: The MDA-MB-231 cells morphology after treatment with BODIPY **2b**; Figure S22. The viability of MDA-MB-231 cells after treatment with BODIPYs **1**, **2a**, and **2b** under normoxic conditions; Figure S23. The viability of MDA-MB-231 cells after treatment with BODIPYs **1**, **2a**, and **2b** under hypoxic conditions; Figure S24. The viability of A2780 cells after treatment with BODIPYs **1**, **2a**, and **2b** under normoxic conditions; Figure S25. The viability of A2780 cells after treatment with BODIPYs **1**, **2a**, and **2b** under hypoxic conditions; Table S1: NMR data for compound **1**, including key correlations determined from ^{1}H-^{1}H COSY, ^{1}H-^{13}C HSQC, and ^{1}H-^{13}C HMBC spectra; Table S2: NMR data for BODIPY **2a**, including key correlations determined from ^{1}H-^{1}H COSY, ^{1}H-^{13}C HSQC, and ^{1}H-^{13}C HMBC spectra; Table S3: NMR data for **2b**, including key correlations determined from ^{1}H-^{1}H COSY, ^{1}H-^{13}C HSQC, and ^{1}H-^{13}C HMBC spectra; Table S4: UV-Vis absorption maxima (λ_{Abs}) and logarithms of molar absorption coefficients (log ε) of compounds **1**, **2a**, and **2b** in various solvents; Table S5: The cell viability of A2780 and MDA-MB-231 treated with **2a** and **2b** under hypoxic conditions without exposure to light; Table S6:

The cell viability of A2780 and MDA-MB-231 treated with **2a** and **2b** under hypoxic conditions after irradiation at a light dose of 2 J/cm^2.

Author Contributions: Conceptualization, J.P. and M.K.; methodology. J.P. and M.K.; validation, W.P. and M.K.; formal analysis, W.P., J.P. and M.K.; investigation, W.P., N.K. and M.K.; resources, W.P., M.M., M.K. and J.P.; data curation, W.P., J.P. and M.K.; writing—original draft preparation, W.P., J.P. and M.K.; writing—review and editing, M.M.; visualization, W.P., J.P. and M.K.; supervision, J.P., M.K. and M.M; funding acquisition, W.P., M.M. and M.K. All authors have read and agreed to the published version of the manuscript.

Funding: This study was supported by the National Science Centre, Poland, under grant No. 2021/41/N/NZ7/00371.

Institutional Review Board Statement: Not applicable.

Informed Consent Statement: Not applicable.

Data Availability Statement: Data are contained within the article/Supplementary Material, further inquiries can be directed to the corresponding authors.

Acknowledgments: The graphical abstract was prepared using BioRender.com (under the appropriate license, license holder M.M.; last accessed on 18 July 2024).

Conflicts of Interest: The authors declare no conflicts of interest.

References

1. Zhang, D.; Liu, L.; Zhang, X.; Lu, J.; Jiang, X.-D. Rational Design of Photofunctional Dyes BODIPYs/Aza-BODIPYs and Applications for Photocatalysis, Photoelectric Conversion and Thermochromic Materials. *Resour. Chem. Mater.* **2024**, *3*, 103–122. [CrossRef]
2. Ilhan, H.; Cakmak, Y. Functionalization of BODIPY Dyes with Additional C–N Double Bonds and Their Applications. *Top. Curr. Chem.* **2023**, *381*, 28. [CrossRef] [PubMed]
3. Das, S.; Dey, S.; Patra, S.; Bera, A.; Ghosh, T.; Prasad, B.; Sayala, K.D.; Maji, K.; Bedi, A.; Debnath, S. BODIPY-Based Molecules for Biomedical Applications. *Biomolecules* **2023**, *13*, 1723. [CrossRef] [PubMed]
4. Zhang, S.; Qu, Y.; Zhang, D.; Li, S.; Tang, F.; Ding, A.; Hu, L.; Zhang, J.; Wang, H.; Huang, K.; et al. Rational Design and Biological Application of Hybrid Fluorophores. *Chem. Eur. J.* **2024**, *30*, e202303208. [CrossRef] [PubMed]
5. Cheng, H.; Cao, X.; Zhang, S.; Zhang, K.; Cheng, Y.; Wang, J.; Zhao, J.; Zhou, L.; Liang, X.; Yoon, J. BODIPY as a Multifunctional Theranostic Reagent in Biomedicine: Self-Assembly, Properties, and Applications. *Adv. Mater.* **2023**, *35*, 2207546. [CrossRef] [PubMed]
6. Malacarne, M.C.; Gariboldi, M.B.; Caruso, E. BODIPYs in PDT: A Journey through the Most Interesting Molecules Produced in the Last 10 Years. *Int. J. Mol. Sci.* **2022**, *23*, 10198. [CrossRef]
7. Yadav, I.S.; Misra, R. Design, Synthesis and Functionalization of BODIPY Dyes: Applications in Dye-Sensitized Solar Cells (DSSCs) and Photodynamic Therapy (PDT). *J. Mater. Chem. C* **2023**, *11*, 8688–8723. [CrossRef]
8. Lo, P.; Ng, D.K.P.; Pandey, R.K.; Zimcik, P. Photodynamic Therapy: An Innovative and Versatile Treatment Modality Triggered by Light. *ChemPlusChem* **2023**, *88*, e202300159. [CrossRef] [PubMed]
9. Michalak, M.; Mazurkiewicz, S.; Szymczyk, J.; Ziental, D.; Sobotta, Ł. Photodynamic Therapy Applications—Review. *JMS* **2023**, *92*, e865. [CrossRef]
10. Hong, L.; Li, J.; Luo, Y.; Guo, T.; Zhang, C.; Ou, S.; Long, Y.; Hu, Z. Recent Advances in Strategies for Addressing Hypoxia in Tumor Photodynamic Therapy. *Biomolecules* **2022**, *12*, 81. [CrossRef]
11. Udomsak, W.; Kucinska, M.; Pospieszna, J.; Dams-Kozlowska, H.; Chatuphonprasert, W.; Murias, M. Antioxidant Enzymes in Cancer Cells: Their Role in Photodynamic Therapy Resistance and Potential as Targets for Improved Treatment Outcomes. *Int. J. Mol. Sci.* **2024**, *25*, 3164. [CrossRef] [PubMed]
12. Liu, M.; Ma, S.; She, M.; Chen, J.; Wang, Z.; Liu, P.; Zhang, S.; Li, J. Structural Modification of BODIPY: Improve Its Applicability. *Chin. Chem. Lett.* **2019**, *30*, 1815–1824. [CrossRef]
13. Kamkaew, A.; Lim, S.H.; Lee, H.B.; Kiew, L.V.; Chung, L.Y.; Burgess, K. BODIPY Dyes in Photodynamic Therapy. *Chem. Soc. Rev.* **2013**, *42*, 77–88. [CrossRef]
14. Piskorz, J.; Porolnik, W.; Kucinska, M.; Dlugaszewska, J.; Murias, M.; Mielcarek, J. BODIPY-Based Photosensitizers as Potential Anticancer and Antibacterial Agents: Role of the Positive Charge and the Heavy Atom Effect. *ChemMedChem* **2021**, *16*, 399–411. [CrossRef]

15. Loudet, A.; Burgess, K. BODIPY Dyes and Their Derivatives: Syntheses and Spectroscopic Properties. *Chem. Rev.* **2007**, *107*, 4891–4932. [CrossRef]
16. Prieto-Montero, R.; Prieto-Castañeda, A.; Katsumiti, A.; Sola-Llano, R.; Agarrabeitia, A.R.; Cajaraville, M.P.; Ortiz, M.J.; Martinez-Martinez, V. Red haloBODIPYs as Theragnostic Agents: The Role of the Substitution at *Meso* Position. *Dye. Pigment.* **2022**, *198*, 110015. [CrossRef]
17. Hu, W.; Zhang, X.-F.; Lu, X.; Lan, S.; Tian, D.; Li, T.; Wang, L.; Zhao, S.; Feng, M.; Zhang, J. Modifying the Meso -Phenyl with Electron Donating Amino Groups Strongly Enhances BODIPY's Ability as Good Singlet Oxygen Photosensitizer. *Dye. Pigment.* **2018**, *149*, 306–314. [CrossRef]
18. Ko, S.; Kim, C.Y.; Damodar, K.; Lim, H.M.; Kim, J.H.; Lee, C.-H.; Lee, J.T. Substituents Modification of *Meso*-Aryl BODIPYs for Tuning Photophysical Properties. *Tetrahedron* **2018**, *74*, 287–295. [CrossRef]
19. Guseva, G.B.; Antina, E.V.; Berezin, M.B.; Pavelyev, R.S.; Kayumov, A.R.; Sharafutdinov, I.S.; Lisovskaya, S.A.; Lodochnikova, O.A.; Islamov, D.R.; Usachev, K.S.; et al. *Meso*-Substituted-BODIPY Based Fluorescent Biomarker: Spectral Characteristics, Photostability and Possibilities for Practical Application. *J. Photochem. Photobiol. A Chem.* **2020**, *401*, 112783. [CrossRef]
20. Hu, W.; Zhang, X.-F.; Lu, X.; Lan, S.; Tian, D.; Li, T.; Wang, L.; Zhao, S.; Feng, M.; Zhang, J. Attaching Electron Donating Groups on the *Meso*-Phenyl and *Meso*-Naphthyl Make Aryl Substituted BODIPYs Act as Good Photosensitizer for Singlet Oxygen Formation. *J. Lumin.* **2018**, *194*, 185–192. [CrossRef]
21. Shi, W.-J.; Chen, R.; Yang, J.; Wei, Y.-F.; Guo, Y.; Wang, Z.-Z.; Yan, J.; Niu, L. Novel *Meso*-Benzothiazole-Substituted BODIPY-Based AIE Fluorescent Rotor for Imaging Lysosomal Viscosity and Monitoring Autophagy. *Anal. Chem.* **2022**, *94*, 14707–14715. [CrossRef]
22. Liu, R.; Qian, Y. NIR Ditriphenylamine Indole-BODIPY Photosensitizer: Synthesis, Photodynamic Therapy in A549 Cells and Two-Photon Fluorescence Imaging in Zebrafish. *Spectrochim. Acta Part A Mol. Biomol. Spectrosc.* **2024**, *304*, 123387. [CrossRef] [PubMed]
23. Gonçalves, R.C.R.; Teixeira, F.; Peñalver, P.; Costa, S.P.G.; Morales, J.C.; Raposo, M.M.M. Designing Antitrypanosomal and Antileishmanial BODIPY Derivatives: A Computational and In Vitro Assessment. *Molecules* **2024**, *29*, 2072. [CrossRef] [PubMed]
24. Tursynova, N.; Helena Maliszewska, I.; Jóźwiak, K.; Sokolnicki, J.; Kochel, A.; Lipkowski, P.; Bartkiewicz, S.; Filarowski, A. The Photoinactivation of Pathogenic Bacteria Using Synthesized Benzodioxole-BODIPY Dyes. *J. Photochem. Photobiol. A Chem.* **2024**, *450*, 115474. [CrossRef]
25. Mancini, R.S.; Barden, C.J.; Weaver, D.F.; Reed, M.A. Furazans in Medicinal Chemistry. *J. Med. Chem.* **2021**, *64*, 1786–1815. [CrossRef]
26. Verdoliva, V.; Digilio, G.; Saviano, M.; De Luca, S. Thio-Conjugation of Substituted Benzofurazans to Peptides: Molecular Sieves Catalyze Nucleophilic Attack on Unsaturated Fused Rings. *Catal. Sci. Technol.* **2021**, *11*, 1067–1076. [CrossRef]
27. Li, L.; Kracht, J.; Peng, S.; Bernhardt, G.; Buschauer, A. Synthesis and Pharmacological Activity of Fluorescent Histamine H1 Receptor Antagonists Related to Mepyramine. *Bioorg. Med. Chem. Lett.* **2003**, *13*, 1245–1248. [CrossRef]
28. Kang, J.; Huo, F.; Ning, P.; Meng, X.; Chao, J.; Yin, C. Two Red-Emission Single and Double 'Arms' Fluorescent Materials Stemed from 'One-Pot' Reaction for Hydrogen Sulfide Vivo Imaging. *Sens. Actuators B Chem.* **2017**, *250*, 342–350. [CrossRef]
29. Ding, X.; Wang, Q.; Chen, D.; Chen, Y.; Pan, W.; Sun, Q.; Chen, Q.; Han, X. A Fluorescent Probe Based on BODIPY for Hydrogen Sulfide Imaging in Living Cells and Zebrafish. *Adv. Agrochem.* **2023**, *2*, 364–370. [CrossRef]
30. Xiang, H.-J.; Tham, H.P.; Nguyen, M.D.; Fiona Phua, S.Z.; Lim, W.Q.; Liu, J.-G.; Zhao, Y. An Aza-BODIPY Based near-Infrared Fluorescent Probe for Sensitive Discrimination of Cysteine/Homocysteine and Glutathione in Living Cells. *Chem. Commun.* **2017**, *53*, 5220–5223. [CrossRef]
31. Parisi, C.; Longobardi, G.; Graziano, A.C.E.; Fraix, A.; Conte, C.; Quaglia, F.; Sortino, S. A Molecular Dyad Delivered by Biodegradable Polymeric Nanoparticles for Combined PDT and NO-PDT in Cancer Cells. *Bioorg. Chem.* **2022**, *128*, 106050. [CrossRef]
32. Bumagina, N.A.; Antina, E.V.; Ksenofontov, A.A.; Antina, L.A.; Kalyagin, A.A.; Berezin, M.B. Basic Structural Modifications for Improving the Practical Properties of BODIPY. *Coord. Chem. Rev.* **2022**, *469*, 214684. [CrossRef]
33. Magde, D.; Wong, R.; Seybold, P.G. Fluorescence Quantum Yields and Their Relation to Lifetimes of Rhodamine 6G and Fluorescein in Nine Solvents: Improved Absolute Standards for Quantum Yields. *Photochem. Photobiol.* **2002**, *75*, 327–334. [CrossRef]
34. Lim, S.H.; Thivierge, C.; Nowak-Sliwinska, P.; Han, J.; van den Bergh, H.; Wagnières, G.; Burgess, K.; Lee, H.B. In Vitro and In Vivo Photocytotoxicity of Boron Dipyrromethene Derivatives for Photodynamic Therapy. *J. Med. Chem.* **2010**, *53*, 2865–2874. [CrossRef] [PubMed]
35. Porolnik, W.; Kasprzycka, M.; Podciechowska, K.; Teubert, A.; Piskorz, J. Synthesis and Spectroscopic Properties of Novel Dipyrrole and Tetrapyrrole-Based Photosensitizers with Various Biphenylyl Substituents. *Tetrahedron* **2022**, *127*, 133088. [CrossRef]
36. Kryjewski, M.; Rebis, T.; Milczarek, G.; Gdaniec, Z.; Goslinski, T.; Mielcarek, J. Magnesium(II) 1-(1-Adamantylsulfanyl)Phthalocy anine—Synthesis, Photochemical and Electrochemical Properties. *New J. Chem.* **2016**, *40*, 9774–9780. [CrossRef]

37. Karstens, T.; Kobs, K. Rhodamine B and Rhodamine 101 as Reference Substances for Fluorescence Quantum Yield Measurements. *J. Phys. Chem.* **1980**, *84*, 1871–1872. [CrossRef]
38. Porolnik, W.; Ratajczak, M.; Mackowiak, A.; Murias, M.; Kucinska, M.; Piskorz, J. Liposomal Formulations of Novel BODIPY Dimers as Promising Photosensitizers for Antibacterial and Anticancer Treatment. *Molecules* **2024**, *29*, 5304. [CrossRef] [PubMed]
39. Lagorio, M.G.; Dicelio, L.E.; San Roman, E.A.; Braslavsky, S.E. Quantum Yield of Singlet Molecular Oxygen Sensitization by Copper(II) Tetracarboxyphthalocyanine. *J. Photochem. Photobiol. B Biol.* **1989**, *3*, 615–624. [CrossRef]
40. Kochevar, I.E.; Redmond, R.W. [2] Photosensitized Production of Singlet Oxygen. In *Methods in Enzymology*; Elsevier: Amsterdam, The Netherlands, 2000; Volume 319, pp. 20–28. ISBN 978-0-12-182220-0.
41. Kucinska, M.; Skupin-Mrugalska, P.; Szczolko, W.; Sobotta, L.; Sciepura, M.; Tykarska, E.; Wierzchowski, M.; Teubert, A.; Fedoruk-Wyszomirska, A.; Wyszko, E.; et al. Phthalocyanine Derivatives Possessing 2-(Morpholin-4-Yl)Ethoxy Groups as Potential Agents for Photodynamic Therapy. *J. Med. Chem.* **2015**, *58*, 2240–2255. [CrossRef]
42. Piskorz, J.; Dlugaszewska, J.; Porolnik, W.; Teubert, A.; Mielcarek, J. Boron-Dipyrromethene Derivatives Bearing N-Alkyl Phthalimide and Amine Substituents of Potential Application in the Photoinactivation of Bacteria. *Dye. Pigment.* **2020**, *178*, 108322. [CrossRef]
43. Galeotti, F.; Calabrese, V.; Cavazzini, M.; Quici, S.; Poleunis, C.; Yunus, S.; Bolognesi, A. Self-Functionalizing Polymer Film Surfaces Assisted by Specific Polystyrene End-Tagging. *Chem. Mater.* **2010**, *22*, 2764–2769. [CrossRef]
44. Krzemien, W.; Rohlickova, M.; Machacek, M.; Novakova, V.; Piskorz, J.; Zimcik, P. Tuning Photodynamic Properties of BODIPY Dyes, Porphyrins' Little Sisters. *Molecules* **2021**, *26*, 4194. [CrossRef] [PubMed]
45. Shen, B.; Qian, Y.; Qi, Z.; Lu, C.; Sun, Q.; Xia, X.; Cui, Y. Near-Infrared BODIPY-Based Two-Photon ClO^- Probe Based on Thiosemicarbazide Desulfurization Reaction: Naked-Eye Detection and Mitochondrial Imaging. *J. Mater. Chem. B* **2017**, *5*, 5854–5861. [CrossRef]
46. Banfi, S.; Caruso, E.; Zaza, S.; Mancini, M.; Gariboldi, M.B.; Monti, E. Synthesis and Photodynamic Activity of a Panel of BODIPY Dyes. *J. Photochem. Photobiol. B Biol.* **2012**, *114*, 52–60. [CrossRef] [PubMed]
47. Wang, J.; Gong, Q.; Wang, L.; Hao, E.; Jiao, L. The Main Strategies for Tuning BODIPY Fluorophores into Photosensitizers. *J. Porphyr. Phthalocyanines* **2020**, *24*, 603–635. [CrossRef]
48. May, A.K.; Chiyumba, C.; Harris, J.; Mack, J.; Nyokong, T. Photodynamic Antimicrobial Activities of Halogenated 3,5-Dimethyl- and 1,3,5,7-Tetramethyl-*Meso*-Pentafluorophenyl BODIPY Dyes. *J. Porphyr. Phthalocyanines* **2022**, *26*, 691–700. [CrossRef]
49. Çınar, H.Ş.; Özçelik, Ş.; Kaya, K.; Kutlu, Ö.D.; Erdoğmuş, A.; Gül, A. Synthesis and Photophysical Properties of Monomeric and Dimeric Halogenated Aza-BODIPYs. *J. Mol. Struct.* **2020**, *1200*, 127108. [CrossRef]
50. Turksoy, A.; Yildiz, D.; Akkaya, E.U. Photosensitization and Controlled Photosensitization with BODIPY Dyes. *Coord. Chem. Rev.* **2019**, *379*, 47–64. [CrossRef]
51. Caruso, E.; Malacarne, M.C.; Marras, E.; Papa, E.; Bertato, L.; Banfi, S.; Gariboldi, M.B. New BODIPYs for Photodynamic Therapy (PDT): Synthesis and Activity on Human Cancer Cell Lines. *Bioorg. Med. Chem.* **2020**, *28*, 115737. [CrossRef] [PubMed]
52. Caruso, E.; Gariboldi, M.; Sangion, A.; Gramatica, P.; Banfi, S. Synthesis, Photodynamic Activity, and Quantitative Structure-Activity Relationship Modelling of a Series of BODIPYs. *J. Photochem. Photobiol. B Biol.* **2017**, *167*, 269–281. [CrossRef]
53. Badon, I.W.; Jee, J.-P.; Vales, T.P.; Kim, C.; Lee, S.; Yang, J.; Yang, S.K.; Kim, H.-J. Cationic BODIPY Photosensitizers for Mitochondrion-Targeted Fluorescence Cell-Imaging and Photodynamic Therapy. *Pharmaceutics* **2023**, *15*, 1512. [CrossRef] [PubMed]

biomedicines

Article

Predictive Factors for Morphological and Functional Improvements in Long-Lasting Central Serous Chorioretinopathy Treated with Photodynamic Therapy

Maciej Gawęcki [1,2,*], Krzysztof Kiciński [1,2], Jan Kucharczuk [3], Monika Gołębiowska-Bogaj [1] and Andrzej Grzybowski [4,5]

1 Department of Ophthalmology, Pomeranian Hospitals, 84-200 Wejherowo, Poland; krzysztofkg999@icloud.com (K.K.); monikaxvz@gmail.com (M.G.-B.)
2 Dobry Wzrok Ophthalmological Center, 80-392 Gdansk, Poland
3 Department of Ophthalmology, 10th Military Research Hospital and Polyclinic, 85-681 Bydgoszcz, Poland; jankucharczuk@wp.pl
4 Department of Ophthalmology, University of Warmia and Mazury, Oczapowskiego 2, 10-719 Olsztyn, Poland; ae.grzybowski@gmail.com
5 Institute for Research in Ophthalmology, Foundation for Ophthalmology Development, Mickiewicza 24, 61-836 Poznan, Poland
* Correspondence: maciej@gawecki.com

Abstract: Backgrounds: Photodynamic therapy (PDT) is an established treatment modality in central serous chorioretinopathy (CSCR). The goal of our study was to evaluate the morphological and functional effects of PDT in patients with long-lasting CSCR and determine the related predictive factors for improvement. **Methods**: This retrospective analysis included consecutive patients with chronic CSCR who consented to PDT. The material comprised 98 eyes of 81 patients (67 males and 14 females) with a disease duration longer than 6 months followed for 6 months post treatment. All patients underwent a basic ophthalmological examination including best corrected visual acuity (BCVA) testing and imaging, spectral-domain optical coherence tomography (SD-OCT), and fluorescein angiography. Patients without macular neovascularization (MNV) were subjected to half-dose PDT (3 mg/m^2) with standard fluence (50 J/cm^2), guided by indocyanine green angiography. Cases complicated by MNV were subjected to full-dose PDT. **Results**: A morphological response, defined as complete resolution of subretinal fluid, was achieved in 76.29% of cases, and an improvement in BCVA of at least one logMAR line was obtained in 77.53% of cases. The mean BCVA gain was 1.2 logMAR line. All SD-OCT measurements (central retinal thickness, macular volume, mean subfield thickness, subretinal fluid height, and subfoveal choroidal thickness) showed a significant reduction post PDT. A multivariate analysis proved better morphological outcome associations with a younger age and male gender and better visual gains achieved in patients without intraretinal abnormalities. Univariate testing also showed strong relationships between better baseline BCVA and greater functional and morphological improvements, between shorter disease duration and morphological gains, and between the absence of MNV or intraretinal abnormalities and morphological gains. PDT was highly effective in providing a resolution of pigment epithelial detachment ($p = 0.0004$). The observed effect was significantly dependent upon the lower baseline central retinal thickness ($p = 0.0095$). Patients with intraretinal abnormalities or MNV showed moderate improvements post PDT. **Conclusions**: PDT in long-lasting CSCR cases provides good morphological results but generally minor visual gains. Patients' expectations of significant increases in BCVA after prolonged disease with distinct alterations of the neurosensory retina should be managed.

 https://doi.org/10.3390/biomedicines13040944

Keywords: central serous chorioretinopathy; photodynamic therapy; spectral-domain optical coherence tomography; best corrected visual acuity; subretinal fluid

1. Introduction

Central serous chorioretinopathy (CSCR) is a quite common clinical entity affecting predominantly young individuals, usually in their fourth to sixth decades of life [1,2]. CSCR's pathogenesis is a subject of debate and so are the recommendations for its treatment, so far lacking an international consensus [3]. The chronic form of CSCR, developing in about 85% of cases, is a distinct therapeutic problem as it leads to significant visual loss [4,5]. Among the most effective treatment modalities tried in this form of CSCR are classic laser photocoagulation (CLP) of the leakage point, photodynamic therapy (PDT), subthreshold micropulse laser (SML), and systemic mineralocorticoid inhibitors, such as Eplerenon. The use of CLP is limited only to cases with well-defined leakage points located outside the fovea; thus, it cannot be applied in a significant number of affected patients [6,7]. SML is a cheap treatment modality that proves effective in 40–80% of cases; however, it does not prevent disease recurrences [8]. Mineralocorticoid inhibitors have been widely used for chronic CSCR management in the last decade, with variable success. A few metanalyses showed that their efficacy in providing remission from symptoms is not superior to that of observation or other treatment options [9–11]. Recent studies advocate the superiority of PDT over other treatment modalities in CSCR [12,13]. PDT has been used to treat CSCR for more than 20 years [14–16]. Although its application has been tested in both acute and chronic cases, it is usually applied in long-lasting CSCR because of its high cost and invasive character. The mechanism of action of PDT is still unclear. The most popular pathogenic theory assumes that transient hypoperfusion at the choriocapillaris, elicited by the stimulation of the photosensitizer verteporfin, results in decreased choroidal congestion and hyperpermeability of the choriocapillaris, enabling the absorption of subretinal and intraretinal fluids [17–19]. This mechanism aligns with the pachychoroid theory of CSCR's etiology, which assumes a significantly increased choroidal thickness in CSCR patients [20,21]. PDT with different doses of verteporfin and laser fluences has been employed to treat CSCR—a full dose (6 mg/m^2), half dose (3 mg/m^2), standard fluence (50 J/cm^2), or reduced fluence (25 J/cm^2)—with similar results. The reported efficacy of PDT in chronic CSCR is considerable in terms of the improvement in retinal morphology and moderate in terms of functional gain [22–24]. These results suggest that baseline factors should be considered when predicting morphological and functional improvements after PDT.

The goal of our study was to evaluate the morphological and functional effects of PDT in patients with long-lasting CSCR and to identify the predictive factors associated with patient improvement. These factors included patient demographics, the disease duration, and morphological and functional features obtained from visual acuity testing and spectral-domain optical coherence tomography (SD-OCT) scanning.

2. Materials and Methods

This study was approved by the local bioethical commission of Okręgowa Izba Lekarska (approval number KB-35/2023, dated 16 August 2023). All procedures performed during the study complied with the Declaration of Helsinki.

The retrospective analysis included all consecutive Caucasian patients with chronic CSCR who consented to PDT performed at Dobry Wzrok Ophthalmological Center between June 2022 and October 2024.

The material comprised 98 eyes of 81 patients (67 males and 14 females). All patients were questioned about the duration of their symptoms, and their medical history, including retinal imaging, was carefully reviewed. The study included active CSCR cases that had lasted longer than 6 months and/or had a recurrent course. Before the PDT, most patients had been unsuccessfully treated with classic photocoagulation or subthreshold micropulse laser, anti-VEGF intravitreal injections, and topical or systemic nonsteroidal anti-inflammatory medications.

We adopted the diagnostic criteria for CSCR from the report of the Central Serous Chorioretinopathy International Group [25]. These criteria are based on the results of multimodal imaging, with major criteria including past or present subretinal fluid (SRF) detected on SD-OCT scans, leakage observed on fluorescein angiography (FA), alteration of the retinal pigment epithelium (RPE) visible on fundus autofluorescence (FAF), areas of increased permeability of the choriocapillaris observed on indocyanine green angiography (ICGA), and increased choroidal thickness on SD-OCT scans.

At baseline, all patients underwent a basic ophthalmological examination, including best corrected visual acuity (BCVA) testing, intraocular pressure measurement, slit lamp examination of the anterior segment and eye fundus, SD-OCT scanning, OCT angiography, FAF, and FA, performed either at the center or by the referring practitioner (Visucam 524, 2019, Carl Zeiss Meditec AG, Jena, Germany). ICGA was always performed with the same device in inconclusive cases and before the PDT. The presence of macular neovascularization (MNV) was determined by ICGA and angio-OCT. All procedures except FA and ICGA were also performed at the two follow-up visits: one month and six months post PDT.

SD-OCT measurements were performed with a REVO FC 130 system (2023, Optopol Technology, Zawiercie, Poland). They included the central subfoveal thickness (CST), corresponding to the mean retinal thickness within a central circle of 1 mm diameter; mean subfield thickness (MST), corresponding to the mean retinal thickness in a central circle of 6 mm diameter; macular volume (MV), which is the retinal volume over a central circle of 6 mm diameter in the central RPE plan; maximum subretinal fluid (SRF) height, measured manually; and subfoveal choroidal thickness (SFCT), also measured manually under the foveola. Additionally, the presence of the morphological features of pigment epithelial detachment (PED), neurosensory retina abnormalities (NRAs), including mainly intraretinal cysts, and MNV was also noted for each case before and after treatment.

In patients with CSCR but without MNV, PDT was performed with a half dose of verteporfin (3 mg/m^2) and a standard laser fluence of 50 J/cm^2 (Vitra 689, 2020, Quantel Medical, Cournon d'Auvergne, France). Patients with MNV had a full dose of verteporfin (6 mg/m^2) with standard laser fluence applied during the procedure. The amount of verteporfin required for each patient was calculated from their mass and height. The range of irradiation performed with a 689 nm laser was determined by ICGA examination. The laser spot size was adjusted to cover the entire area of choroidal hyperpermeability determined on the photographs obtained from the ICGA fundus camera with a 500 μm margin. The mean value of the spot size employed in the study cohort was 4361 +/− 1122 μm, with median of 4500 μm and quartiles Q_1–Q_3 = 4000–4500 μm. In cases with multiple targets, a second irradiation was performed immediately after the first one. The duration of a single irradiation was 83 s, according to standards of the PDT procedure. The same protocol was applied in all patients.

In this study, patients were categorized as responders and non-responders to PDT according to their retinal morphology and visual function. A patient was considered a morphological responder if complete resorption of SRF was observed after PDT, whereas partial resolution of SRF was considered a non-response. A functional responder was defined as a patient whose BCVA improved by at least one line on the logMAR chart after

PDT. For the purpose of the statistical analysis, categorization of responders at six months post PDT was considered.

2.1. Statistical Analysis

The results of the treatment of chronic CSCR with PDT were analyzed in terms of the changes in the BCVA and SD-OCT parameters after treatment. The groups of responders and non-responders were then compared in terms of their baseline BCVA, SD-OCT measurements, and presence or absence of PED, NRA, and MNV. Both multivariate and univariate analyses were performed to assess the impact of different factors on morphological and functional outcomes of the treatment.

2.2. Statistical Procedures

Categorical variables (frequencies) are described as integers and percentages. Measurements are expressed as means, medians, standard deviations, and lower and upper quartiles. The normality of distributions was tested by the Shapiro–Wilk W test. Levene's test was performed to estimate the homogeneity of variances. A multifactor analysis of variance (ANOVA) with repeated measurements was conducted to assess the dynamics of normally distributed numerical variables and their differences between the study groups. General estimating equations were applied for non-normally distributed variables. A binary logistic regression model was fitted in order to estimate the odds ratios and their significance for dichotomous explained variables. All multivariate procedures were controlled for the participants' age, gender, disease duration, and comorbidities. A level of $p < 0.05$ was considered statistically significant. All procedures were carried out using Statistica™ release 13.3 (TIBCO Software Inc., Palo Alto, CA, USA).

3. Results

The mean age of the patients was 48.78 ± 10.22 years (Me = 47, Q_1–Q_3 =43–54) and the mean disease duration was 63.35 ± 75.31 months (Me = 48, Q_1–Q_3 = 20–100). The BCVA and morphological parameters measured by SD-OCT significantly improved during the 6-month follow-up period (Table 1). The effect of SRF height and CST reduction after the PDT was enhanced after the first follow-up visit at 1 month and proved greater at 6 months of follow-up. The BCVA improvement at 6 months was no different from the functional effect at 1 month post treatment. The relationships between the occurrence of morphological and functional effects and baseline qualitative and quantitative features are provided in Tables 2 and 3. Generally, morphological and functional improvements were noted in similar percentages of patients. Morphological improvements were noted significantly more frequently in male patients in both univariate and multivariate analyses. Functional improvements were poorer in cases with neurosensory retina abnormalities for both of these statistical tests. The univariate analysis showed poorer morphological responses in patients with NRAs or MNV as well. An analysis of morphological and functional responses according to baseline BCVA showed significant differences in the univariate analysis: patients with better baseline visual acuity responded better to treatment and had a tendency for higher visual gains (Table 2). Specifically, BCVA > 0.6 logMAR showed similar percentage of morphological responses and failures, while only one eye with BCVA ≤ 0.2 logMAR did not show any improvement post PDT. Morphological improvements were also strongly associated with a younger age, as proved in both univariate and multivariate tests. The univariate analysis showed higher percentages of morphological resolution in patients with shorter disease durations.

Table 1. Descriptive statistics for treatment-related changes in numerical traits over the 6-month study period (n = 98 eyes).

Analyzed Trait	Follow-Up	Statistical Parameter *				*p* Value **
		M	*SD*	*Me*	Q_1–Q_3	
BCVA [logMAR]	Before treatment	0.46	0.37	0.40	0.20–0.60	<0.0001
	At 1 month	0.35	0.34	0.30	0.10–0.50	0.6547
	At 6 months	0.34	0.38	0.20	0.10–0.50	<0.0001
CST [μm]	Before treatment	327.26	82.06	307.00	269.00–396.00	<0.0001
	At 1 month	249.41	80.63	226.00	194.00–275.00	<0.0001
	At 6 months	244.10	76.34	227.00	195.00–264.00	<0.0001
MST [μm]	Before treatment	302.82	29.46	298.00	283.00–312.00	<0.0001
	At 1 month	281.79	26.89	277.00	267.00–294.00	0.1589
	At 6 months	284.49	29.69	280.00	268.00–297.00	<0.0001
MV [mm^3]	Before treatment	8.54	0.81	8.43	8.01–8.79	<0.0001
	At 1 month	7.97	0.77	7.82	7.54–8.27	0.0779
	At 6 months	8.08	0.90	7.99	7.56–8.47	<0.0001
SFCT [μm]	Before treatment	587.32	110.45	598.00	520.00–660.00	<0.0001
	At 1 month	539.83	102.67	550.00	474.00–603.00	0.8605
	At 6 months	539.09	93.35	548.00	489.00–594.00	<0.0001
SRF, height [μm]	Before treatment	128.26	91.09	120.00	66.00–180.00	<0.0001
	At 1 month	36.95	84.12	0.00	0.00–0.00	0.0184
	At 6 months	30.80	77.76	0.00	0.00–0.00	<0.0001

* Statistical parameters used: M—mean; SD—standard deviation; Me—median; Q—quartile. ** The first *p* value refers to the change observed after 1 month, the second *p* value refers to the change between 1 and 6 months, and the third *p* value concerns the changes over the entire 6-month study period. The models were controlled for the subjects' age, gender, disease duration, and comorbidities. Abbreviations: BCVA—best corrected visual acuity; CST—central subfield thickness; MST—mean subfield thickness; MV—macular volume; SFCT—subfoveal choroidal thickness; SRF—subretinal fluid. Missing data were case-wise deleted.

Table 2. Baseline characteristics of the study cohort by occurrence of improvement observed 6 months after PDT (n = 81 individuals = 98 eyes) (***qualitative traits***).

Analyzed Trait	Morphological Improvement		Functional Improvement		*p* Value *	
	Yes	No	Yes	No	*Univariate*	*Multivariate*
No. of eyes, *n* (%)	74 (76.29)	23 (23.71)	69 (77.53)	20 (22.47)	>0.9999	n/a
Gender, *n* (%):						
• Female	9 (12.16)	8 (34.78)	10 (14.49)	5 (25.00)	0.0127 0.2691	0.0097 OR = 0.14 (95% CI: 0.03–0.62) *p* = 0.1549
• Male	65 (87.84)	15 (65.22)	59 (85.51)	15 (75.00)		
MNV, *n* (%):						
• Present	7 (9.46)	6 (26.09)	7 (10.14)	4 (20.00)	0.0409 0.2792	0.8063 0.9306
• Absent	67 (90.54)	17 (73.91)	62 (89.86)	16 (80.00)		

Table 2. *Cont.*

Analyzed Trait	Morphological Improvement		Functional Improvement		*p* Value *	
	Yes	No	Yes	No	*Univariate*	*Multivariate*
PED, *n* (%):						
• Present	14 (18.92)	5 (21.74)	17 (24.64)	2 (10.00)	0.7660 0.2212	0.3554 0.2122
• Absent	60 (81.08)	18 (78.26)	52 (75.36)	18 (90.00)		
NRA, *n* (%):						
• Present	8 (10.81)	9 (39.13)	8 (11.59)	8 (40.00)	0.0018 0.0036	0.1547 0.0139 OR = 0.16 (95% CI: 0.04–0.069)
• Absent	66 (89.19)	14 (60.87)	61 (88.41)	12 (60.00)		
Baseline BCVA, *n* (%):						
• logMAR ≤ 0.2	24 (32.43)	1 (4.35)	24 (34.78)	1 (5.00)	0.0092 ** 0.0303	0.2747 0.1671
• 0.2 < logMAR ≤ 0.6	43 (58.11)	16 (69.56)	36 (52.18)	16 (80.00)		
• logMAR > 0.6	7 (9.46)	6 (26.09)	9 (13.04)	3 (15.00)		

* The first row: *p* value for morphological improvement; the second row: *p* value for functional improvement. A multivariate binary logistic regression model. ** Paired comparisons: morphological improvement observed when 1 vs. 3 $p = 0.0093$, 1 vs. 2 $p = 0.0392$, 2 vs. 3 $p = 0.1844$; functional improvement observed when 1 vs. 3 $p = 0.0881$, 1 vs. 2 $p = 0.0261$, 2 vs. 3 $p = 0.6940$. Abbreviations: n—integer number, %—percentage, MNV—macular neovascularization; PED—pigment epithelial detachment; NRA—neurosensory retina abnormalities. Missing data were case-wise deleted.

Table 3. Baseline characteristics of the study cohort by occurrence of improvement observed 6 months after PDT (n = 81 individuals = 98 eyes) (***quantitative traits***).

Analyzed Trait	Morphological Improvement		Functional Improvement		*p* Value *	
	Yes	No	Yes	No	*Univariate*	*Multivariate*
Age [year], M (SD); Me (Q_1–Q_3)	46.63 (8.56); 46 (42–52)	55.52 (12.41); 54 (47–66)	48.49 (8.89); 46 (43–54)	53.15 (12.73); 51 (44–62)	0.0012 0.0824	0.0147 OR = 1.08 (95% CI: 1.02–1.15) 0.3319
Disease duration [month], M (SD); Me (Q_1–Q_3)	58.57 (72.30); 30 (20–70)	99.65 (74.87); 96 (48–100)	68.61 (76.79); 48 (20–100)	75.50 (74.84); 60 (27–100)	0.0006 0.3193	0.7329 0.4597
CST [μm], M (SD); Me (Q_1–Q_3)	325.06 (79.12); 306.00 (267–401)	334.30 (92.45); 321 (270–396)	323.03 (84.15); 300 (267–377)	334.15 (76.94); 319 (262–403)	0.6397 0.5975	0.8161 0.9125
MST [μm], M (SD); Me (Q_1–Q_3)	302.58 (27.27); 299 (285–312)	303.61 (36.32); 297 (273–314)	301.81 (29.19); 297 (285–309)	301.35 (31.06); 303 (273–324)	0.8847 0.9512	0.9153 0.6127
Baseline BCVA [logMAR], M (SD); Me (Q_1–Q_3)	0.39 (0.31); 0.35 (0.20–0.50)	0.67 (0,46); 0.60 (0.40–0.80)	0.43 (0.36); 0.40 (0.20–0.60)	0.58 (0.40); 0.50 (0.40–0.60)	0.0008 0.0323	0.0899 0.9566

* The first row: *p* value for morphological improvement; the second row: *p* value for functional improvement. A multivariate binary logistic regression model. Abbreviations: n—integer number, %—percentage, M—mean, SD—standard deviation, Me—median, Q—quartile; MNV—macular neovascularization; PED—pigment epithelial detachment; NRAs—neurosensory retina abnormalities. Missing data were case-wise deleted.

An improvement in the BCVA was noted in 69 eyes (77.53%), with 35 eyes (39.33%) improving by ≤0.1 logMAR and 34 eyes (38.20%) by more than 0.1 logMAR. BCVA stabilization was noted in 11 cases (12.36%), and deterioration of vision occurred in 9 eyes (10.11%).

The BCVA gain, over the 6-month studio period, was meaningfully and positively related to its baseline value ($p = 0.0072$). The multifactor model used did not show any other statistically significant relationships with age ($p = 0.6286$), disease duration ($p = 0.3276$), choroidal thickness ($p = 0.6990$), or retinal thickness ($p = 0.6299$).

The multivariate analysis results for changes in PED and NRAs during the six-month follow-up are provided in Table 4. The prevalence of PED significantly decreased over the

6-month study period ($p = 0.0004$). The multivariate model revealed that the observed effect was statistically significantly dependent upon the baseline CST ($p = 0.0095$): the smaller the CST at baseline, the more pronounced the improvement at 6 months post PDT.

Table 4. Changes in the prevalence of PED and NRAs due to PDT over the 6-month study period (n = 98 eyes) in a multivariate analysis.

Morphological Feature	Follow-Up	Statistical Parameter *		p Value **
		n	%	
PED	Before treatment	19	19.59	0.0004
	At 1 month	5	5.15	
	At 6 months	4	4.12	
NRA	Before treatment	17	17.53	0.6973
	At 1 month	11	11.34	
	At 6 months	9	9.28	

* Statistical parameters used: n—number; %—percentage. ** The models were controlled for the subjects' age, gender, disease duration, baseline BCVA, CST, and MST. Abbreviations: PDT—photodynamic therapy; PED—pigment epithelial detachment; NRA—neurosensory retina abnormalities. Missing data were case-wise deleted.

The prevalence of NRAs fluctuated yet did not change significantly over the 6-month study period in the multivariate model ($p = 0.6973$).

Figures 1–4 present different responses to PDT in patients with long-term CSCR on the SD-OCT scans.

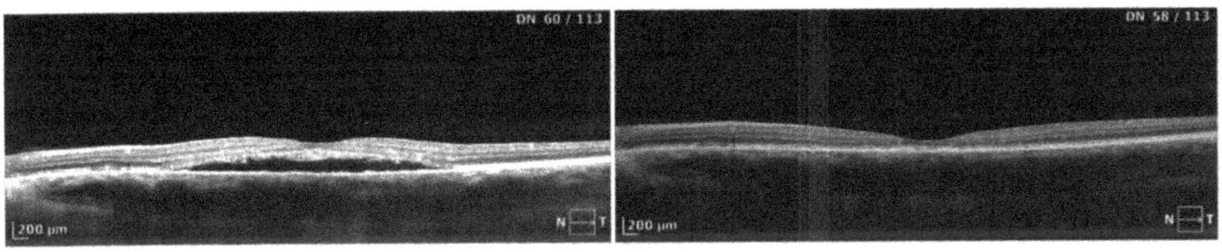

Figure 1. A case of CSCR morphologically presenting with sole subretinal fluid and granulations at the border of the fluid cavity. Complete resolution of the SRF after half dose-PDT was noted at 1 month. Stabilization of the BCVA was noted without improvement at the level of 0.4 logMAR. Significant thinning of the retina can be observed in the macula, which explains the lack of functional gain.

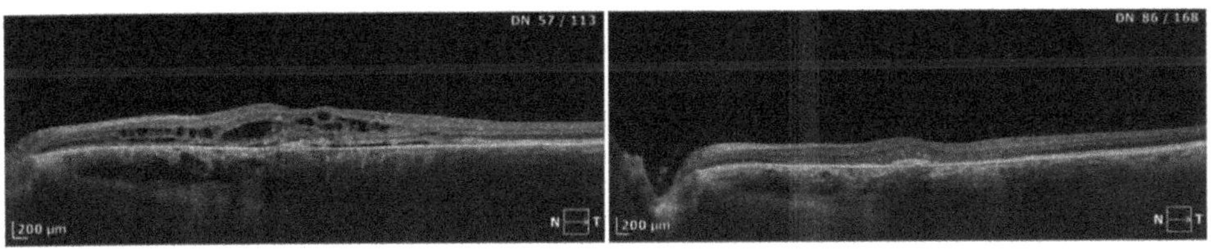

Figure 2. A complex case of CSCR lasting for approximately 36 months, treated with half-dose PDT. Distinct alterations in the neurosensory retina in the form of intraretinal cysts are the predominant morphological feature. A trace of subretinal fluid is also noted. A post-PDT SD-OCT scan revealed considerable resolution of the subretinal and intraretinal fluid. An improvement of 0.1 logMAR was noted post treatment.

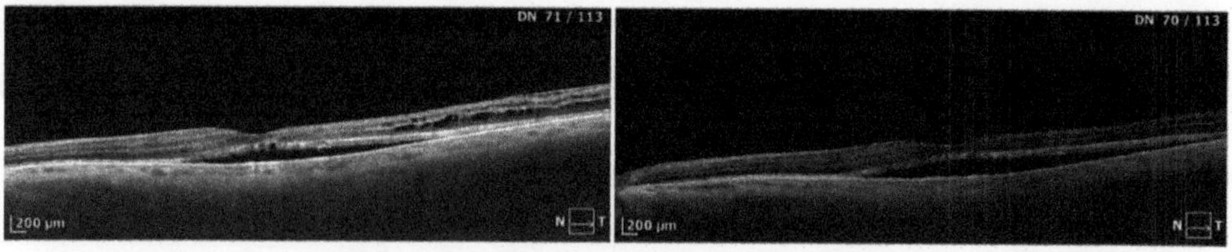

Figure 3. A failure in morphological response after PDT. SD-OCT scans show the case lasting for about 7 years with prominent subretinal fluid and intraretinal cysts. A slight depression of the RPE/choroid line is noted, together with a relatively small choroidal thickness. The amount of SRF was not reduced after PDT, although resolution of the intraretinal cysts was observed. The patient reported a minor improvement in their quality of vision, without gains noted in the visual acuity chart.

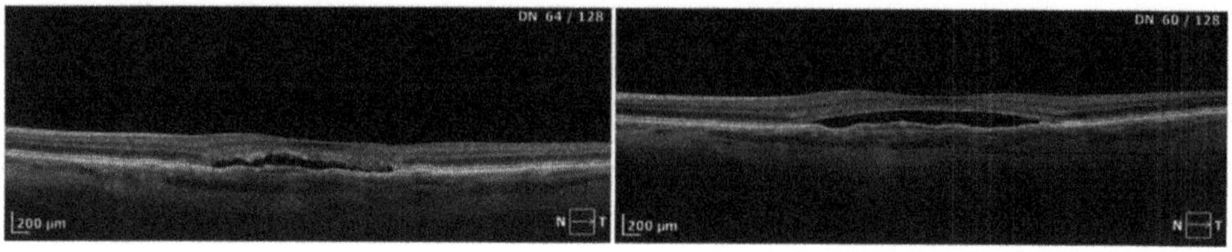

Figure 4. A lack of response to full-dose PDT in a case of chronic CSCR complicated by MNV. At the level of the RPE, characteristic undulation is visible. An increased choroidal thickness was also noted in this case (>400 µm). No reduction in SRF or decrease in the size of sub-RPE MNV was observed. Functional gains were not observed after treatment. The BCVA remained at the level of 0.5 logMAR. The patients remains under close follow-up every 3–4 months. A deterioration in visual acuity was not noted despite the stable morphological picture with MNV and SRF.

4. Discussion

The results of our study indicate the efficacy of PDT in chronic CSCR. Morphological and functional responses were observed in more than 70% of cases. However, the functional gain was generally modest and seldom exceeded one logMAR line. An improved retinal architecture post PDT was observed with the resolution of SRF and reductions in the central subfoveal thickness, mean subfield thickness, and macular volume. Moreover, a significant decrease in choroidal thickness was noted, which, according to the pachychoroid theory, should be the main cause of therapeutic success. Morphological effects were enhanced at six months post PDT compared to one month, with further reductions in the CST and height of SRF. PDT was also effective in providing resolution of PED. This morphological outcome was achieved in most patients and resulted in visual gains. Such a relationship was not true for cases with intraretinal abnormalities or MNV, the presence of which was linked to a poorer response to PDT.

We revealed strong associations between the response to PDT and different baseline factors in a multivariate analysis. Female gender was linked to a poorer morphological response with OR = 0.14; thus, in our cohort, women had 86% smaller odds of achieving the resolution of SRF as compared to the examined men. Poorer functional gain was associated with the presence of neurosensory retina abnormalities (OR = 0.16). These patients had smaller chances of achieving BCVA improvement, by 84%. Lastly, a longer duration of disease effected a worse morphological response (OR = 1.08). The odds of losing such effects were 8% higher with every year of the patient's age.

A univariate analysis also proved other associations of baseline factors with PDT effects, among which baseline visual acuity and disease duration seem to be the most important. Patients with relatively good initial BCVA responded very well to treatment, in both morphological and functional respects. Cases with high BCVA (≤0.2 logMAR) practically always presented with significant improvements. On the other hand, low BCVA (>0.6 logMAR) was associated with a poor morphological response but not necessarily a lack of functional improvement. At this level, patients demonstrated minor visual gains after, for example, a reduction in the CST or intraretinal fluid. Longer durations of symptoms were strongly related to poorer morphological responses post PDT.

The results depicted above have the potential to influence the clinical approach to CSCR cases and enable us to better predict treatment outcomes. Patients burdened with the abovementioned factors should be advised to temper their expectations for perfect outcomes after PDT.

In many cases in this study, the resolution of SRF and the restoration of the retinal architecture did not result in significant functional improvement. The median CST after PDT was only 227.00 μm, compared with the 238 μm reported for healthy subjects by the same group [26]. This finding explains the lack of significant visual recovery observed in many patients. The loss of retinal layers due to the prolonged presence of SRF has also been concluded in many studies [27–30]. It appears that in long-lasting cases, the damage to photoreceptors is sufficient to preclude increased visual acuity.

In the present study, the duration of the disease was long, extending over years rather than months. In this context, the morphological recovery of the retina in the study cohort is highly noteworthy. This is in contrast to the modest functional improvement observed, of about one logMAR line on average. Because the main goal of any ophthalmological treatment is to increase the BCVA, such a result should be considered as moderately successful. It also suggests the importance of early treatment of CSCR, before the loss of retinal cells occurs. In contrast, improvements are less likely to occur after 12 months of CSCR, as reported in a recent study [13].

Interestingly, the morphological response to PDT was poorer in female patients. Note, however, that MNV was present in 5 out of 17 female eyes (almost 30% of the female cohort), compared with 8 out of 81 male eyes (about 10%). Cases of CSCR with MNV seldom respond well to PDT, as observed in other studies [31,32]. The high percentage of eyes with MNV in the female subgroup may partially explain the poor outcome.

Our results should be considered in the context of other reports. Significant morphological recovery after PDT with a minor visual gain has frequently been reported [33–35], with an increase in BCVA exceeding one logMAR line noted in a minority of studies [36,37]. Thus, our results are consistent with those of other reports. Table 5 summarizes the prognostic factors analyzed in recent studies of the response of chronic CSCR to PDT.

Table 5. Factors analyzed in recent studies of the response of chronic CSCR to PDT.

No.	Analyzed Factor	Studies	Finding
1	Disease duration	Wakatsuki et al. 2021 [38], Li et al. 2022 [39], present study	Longer duration associated with poorer morphological outcomes
2	BCVA before PDT	Wakatsuki et al. 2021 [38], van Rijssen et al. 2018 [40], present study	Poorer morphological response at 3 months (Wakatsuki et al.), 2 months (van Rijssen et al.), or 1 month (present study) associated with lower baseline BCVA
3	Age	van Rijssen et al. 2018 [40], Park et al. [37], Arrigo et al. 2021 [41], Nakamura et al. 2022 [42], present study	Older age associated with worse morphological outcomes and non-response
4	Gender	van Rijssen et al. 2018 [40], present study	Anatomical non-response rate greater in male cohort (van Rijssen et al.) or worse anatomical response in female patients (present study)

Table 5. *Cont.*

No.	Analyzed Factor	Studies	Finding
5	Choroidal abnormalities	Arrigo et al. 2021 [41], Nakamura et al. [42]	Poorer morphological response with hyperreflective foci in the choroid (Arrigo et al.), thicker pretreatment choroidal thickness associated with better morphological response (Nakamura et al.)
6	Presence of MNV	Nakamura et al. 2022 [42], Kamimura et al. 2023 [31], present study	Poorer morphological response and/or recurrence in patients with MNV
7	Retinal thickness, condition of retinal layers, intraretinal abnormalities	Wakatsuki et al. 2021 [38], Mirshahi et al. 2024 [43], Li et al. 2022 [39], Sousa et al. 2019 [44], Son et al. 2024 [45], Yu et al. 2022 [46], present study	Higher CST associated with poorer morphological results at 3 months (Wakatsuki et al.), hyperreflective foci in the retina associated with smaller anatomical improvement (Mirshahi et al.), larger SRF and PED at baseline associated with residual SRF post PDT (Li et al.), thicker ONL associated with better morphological (Sousa et al.) and functional (Yu et al.) responses, intact outer retina correlated with better anatomical response (Sousa et al., Son et al.), foveal atrophy associated with poorer morphological and functional outcomes (Son et al.), smaller morphological improvement in patients with intraretinal abnormalities (present study)
8	Leakage in FA	van Rijssen et al. 2018 [40]	Diffuse leakage over area larger than disc diameter associated with poorer morphological outcomes
9	ICGA findings	Jeong et al. 2024 [47], van Rijssen et al. 2018 [40], Ozkaya et al. 2017 [48]	Larger number of macular vortex veins and vortex vein engorgement associated with poorer morphological outcomes and smaller BCVA improvements (Jeong et al.), lack of intense hyperfluorescence on ICGA associated with poorer morphological outcomes (van Rijssen et al.), better anatomical response in patients with midphase focal hyperfluorescence (Ozkaya et al.)
10	Scleral thickness	Forte et al. 2024 [49]	Increased scleral thickness associated with higher risk of anatomical non-responsiveness

Younger age has also been reported as a positive prognostic factor post PDT except in a few studies, with Park et al., van Rijssen et al., and Nakamura et al. noting a better response to PDT in younger patients [37,40,42]. Van Rijssen et al. observed a higher percentage of male non-responders than female non-responders to PDT, in contrast to our results, although their sample size was small [40].

Other authors have identified morphological features not analyzed in this study as predictors of the response to PDT. Mirshahi et al. reported a lower response rate in cases with hyperreflective foci in the retina before treatment [43]. Arrigo et al. analyzed the effect of the same biomarker located in the choroid and obtained a similar outcome [41]. Van Rijssen et al. noted a worse response to PDT in patients with diffuse leakage on FA and without intense hyperfluorescence on ICGA [40]. This finding was confirmed by Ozkaya et al., who reported a better response in patients with midphase hypercyanescence on ICGA [48]. These reports address issues not considered in our study, in which we focused on features that could be reliably quantified, although the scleral thickness was not analyzed.

We did not find a correlation between basic SD-OCT measurements and the response to PDT. However, some studies have reported greater improvements in patients with a thicker outer nuclear layer and an intact external limiting membrane, ellipsoid zone, and interdigitation zone [38,43–46]. These factors were not examined in our study, as we believe that they are difficult to evaluate in the presence of retinal edema, which can bias and distort measurements.

Wakatsuki et al. and van Rijssen et al. also reported an association between a poorer response in patients with a lower baseline BCVA and a longer duration of CSCR, consistent with our results [38,40].

In the context of the published research, our study is noteworthy for its clear definition of non-responsiveness to PDT in both anatomical and functional aspects, as well as its simultaneous analysis of prognostic factors for optimal treatment results in these two areas. In particular, the multivariate analysis provided significant results that could influence current clinical practice.

In contrast, most previous studies only focused on morphological success (dry macula). Such an approach provides only partially reliable conclusions, as many patients exhibiting a perfect anatomical response do not have improved visual function. Moreover, suboptimal morphological responses with reduced intraretinal cysts and reduced SRF can result in minor but observable functional improvements.

Study Limitations

The main study limitation is the lack of a control group. On the other hand, the very long durations of symptoms in the CSCR cases included in this study practically preclude the possibility of spontaneous remission. Thus, we believe that this limitation does not significantly bias the conclusions of the study.

This study focused on SD-OCT measurements for the morphological description of CSCR patients owing to their quantifiability. However, other diagnostic modalities that were not included in our analysis, such as FA and FAF, also provide valuable information on the condition of the RPE.

Another limitation is the relatively small number of female patients in the study cohort, which may have biased the analysis of the outcomes of PDT in female patients. Moreover, our cohort only consisted of Caucasian patients, and an association between response to PDT and race in CSCR was not ruled out in earlier studies [50].

5. Conclusions

The data obtained from this study demonstrated the efficacy of PDT in long-lasting cases of CSCR, although the observed improvements were mainly morphological rather than functional. Patients' expectations of significant increases in BCVA after prolonged disease should be managed. Better morphological response was associated with male gender, shorter disease duration, and younger age, while functional improvements were greater in the absence of intraretinal abnormalities and in cases of better initial visual acuity. SD-OCT measurements of retinal and choroidal thicknesses before treatment cannot solely guide eligibility for PDT in chronic CSCR.

Author Contributions: Conceptualization: M.G. and A.G.; Software: K.K., J.K. and M.G.-B.; Validation: K.K., M.G.-B. and J.K.; Formal analysis: M.G., A.G. and K.K.; Investigation: M.G., K.K. and M.G.-B.; Data curation: K.K., J.K. and M.G.-B.; Writing of original draft: M.G.; Review and editing: A.G., K.K. and J.K. All authors have read and agreed to the published version of the manuscript.

Funding: This research received no external funding.

Institutional Review Board Statement: This study was conducted in accordance with the Declaration of Helsinki and approved by the local bioethical commission of Okręgowa Izba Lekarska (approval no. KB-35/2023, dated 16 August 2023).

Informed Consent Statement: Informed consent was obtained from all subjects involved in the study.

Data Availability Statement: The original contributions presented in this study are included in the article. Further inquiries can be directed to the corresponding author.

Conflicts of Interest: The authors declare no conflicts of interest.

References

1. Daruich, A.; Matet, A.; Dirani, A.; Bousquet, E.; Zhao, M.; Farman, N.; Jaisser, F.; Behar-Cohen, F. Central serous chorioretinopathy: Recent findings and new physiopathology hypothesis. *Prog. Retin. Eye Res.* **2015**, *48*, 82–118. [CrossRef] [PubMed]
2. Ersoz, M.G.; Arf, S.; Hocaoglu, M.; Sayman Muslubas, I.; Karacorlu, M. Patient characteristics and risk factors for central serous chorioretinopathy: An analysis of 811 patients. *Br. J. Ophthalmol.* **2019**, *103*, 725–729. [CrossRef] [PubMed]
3. van Rijssen, T.J.; van Dijk, E.H.C.; Yzer, S.; Ohno-Matsui, K.; Keunen, J.E.E.; Schlingemann, R.O.; Sivaprasad, S.; Querques, G.; Downes, S.M.; Fauser, S.; et al. Central serous chorioretinopathy: Towards an evidence-based treatment guideline. *Prog. Retin. Eye Res.* **2019**, *73*, 100770. [CrossRef] [PubMed]
4. Daruich, A.; Matet, A.; Marchionno, L.; De Azevedo, J.D.; Ambresin, A.; Mantel, I.; Behar-Cohen, F. Acute central serous chorioretinopathy: Factors Influencing Episode Duration. *Retina* **2017**, *37*, 1905–1915. [CrossRef]
5. Gawęcki, M.; Jaszczuk-Maciejewska, A.; Jurska-Jaśko, A.; Kneba, M.; Grzybowski, A. Impairment of visual acuity and retinal morphology following resolved chronic central serous chorioretinopathy. *BMC Ophthalmol.* **2019**, *19*, 160. [CrossRef]
6. Zhou, F.; Yao, J.; Jiang, Q.; Yang, W. Efficacy of Navigated Laser Photocoagulation for Chronic Central Serous Chorioretinopathy: A Retrospective Observational Study. *Dis. Markers* **2022**, *2022*, 7792291. [CrossRef]
7. Robertson, D.M.; Ilstrup, D. Direct, indirect, and sham laser photocoagulation in the management of central serous chorioretinopathy. *Am. J. Ophthalmol.* **1983**, *95*, 457–466. [CrossRef]
8. Gawecki, M.; Pytrus, W.; Swiech, A.; Mackiewicz, J.; Lytvynchuk, L. Laser Treatment of Central Serous Chorioretinopathy—An Update. *Klin. Monbl. Augenheilkd.* **2024**, *241*, 1207–1223. [CrossRef]
9. Salehi, M.; Wenick, A.S.; Law, H.A.; Evans, J.R.; Gehlbach, P. Interventions for central serous chorioretinopathy: A network meta-analysis. *Cochrane Database Syst. Rev.* **2015**, *2015*, CD011841.
10. Feenstra, H.M.A.; van Dijk, E.H.C.; van Rijssen, T.J.; Tsonaka, R.; Diederen, R.M.H.; Schlingemann, R.O.; Hoyng, C.B.; Boon, C.J.F. Crossover to Half-Dose Photodynamic Therapy or Eplerenone in Chronic Central Serous Chorioretinopathy Patients. *Ophthalmol. Retin.* **2022**, *6*, 930–938. [CrossRef]
11. Wang, S.K.; Sun, P.; Tandias, R.M.; Seto, B.K.; Arroyo, J.G. Mineralocorticoid Receptor Antagonists in Central Serous Chorioretinopathy: A Meta-Analysis of Randomized Controlled Trials. *Ophthalmol. Retin.* **2019**, *3*, 154–160. [CrossRef] [PubMed]
12. van Rijssen, T.; van Dijk, E.; Scholz, P.; Breukink, M.; Dijkman, G.; Peters, P.; Tsonaka, R.; MacLaren, R.; Downes, S.; Fauser, S.; et al. Crossover to Photodynamic Therapy or Micropulse Laser After Failure of Primary Treatment of Chronic Central Serous Chorioretinopathy: The REPLACE Trial. *Am. J. Ophthalmol.* **2020**, *216*, 80–89. [CrossRef] [PubMed]
13. Kim, L.A.; Maguire, M.G.; Weng, C.Y.; Smith, J.R.; Jain, N.; Flaxel, C.J.; Patel, S.; Kim, S.J.; Yeh, S. Therapies for Central Serous Chorioretinopathy: A Report by the American Academy of Ophthalmology. *Ophthalmology* **2025**, *132*, 343–353. [CrossRef]
14. Parodi, M.B.; Da Pozzo, S.; Ravalico, G. Photodynamic therapy in chronic central serous chorioretinopathy. *Retina* **2003**, *23*, 235–237. [CrossRef]
15. Yannuzzi, L.A.; Slakter, J.S.; Gross, N.E.; Spaide, R.F.; Costa, D.; Huang, S.J.; Klancnik, J.M., Jr.; Aizman, A. Indocyanine green angiography-guided photodynamic therapy for treatment of chronic central serous chorioretinopathy: A pilot study. *Retina* **2003**, *23*, 288–298. [CrossRef]
16. Iacono, P.; Da Pozzo, S.; Varano, M.; Parravano, M. Photodynamic therapy with verteporfin for chronic central serous chorioretinopathy: A review of data and efficacy. *Pharmaceuticals* **2020**, *13*, 349. [CrossRef]
17. Nassisi, M.; Lavia, C.; Alovisi, C.; Musso, L.; Eandi, C.M. Short-term choriocapillaris changes in patients with central serous chorioretinopathy after half-dose photodynamic therapy. *Int. J. Mol. Sci.* **2017**, *18*, 2468. [CrossRef]
18. Reifeltshammer, E.; Bechstein, L.; Feucht, N.; Lohmann, C.; Maier, M. Effect of low-dose photodynamic therapy at the choriocapillaris level on optical coherence tomography angiography in patients with chronic central serous chorioretinopathy. *Invest. Ophthalmol. Vis. Sci.* **2019**, *60*, 4542.
19. Fernández-Vigo, J.; Moreno-Morillo, F.; Burgos-Blasco, B.; López-Guajardo, L.; Donate-López, J. Early vessel occlusion and recanalization after photodynamic therapy in central serous chorioretinopathy by OCT angiography. *Eur. J. Ophthalmol.* **2021**, *23*, 11206721211027060. [CrossRef]
20. Zhang, X.; Lim, C.Z.F.; Chhablani, J.; Wong, Y.M. Central serous chorioretinopathy: Updates in the pathogenesis, diagnosis and therapeutic strategies. *Eye Vis.* **2023**, *10*, 33. [CrossRef]
21. Kaye, R.; Chandra, S.; Sheth, J.; Boon, C.J.F.; Sivaprasad, S.; Lotery, A. Central serous chorioretinopathy: An update on risk factors, pathophysiology and imaging modalities. *Prog. Retin. Eye Res.* **2020**, *79*, 100865. [CrossRef]
22. Silva, R.M.; Ruiz-Moreno, J.M.; Gomez-Ulla, F.; Montero, J.A.; Gregório, T.; Cachulo, M.L.; Pires, I.A.; Cunha-Vaz, J.G.; Murta, J.N. Photodynamic therapy for chronic central serous chorioretinopathy: A 4-year follow-up study. *Retina* **2013**, *33*, 309–315. [CrossRef] [PubMed]
23. Copete, S.; Ruiz-Moreno, J.M.; Cava, C.; Montero, J.A. Retinal thickness changes following photodynamic therapy in chronic central serous chorioretinopathy. *Graefes Arch. Clin. Exp. Ophthalmol.* **2012**, *250*, 803–808. [CrossRef] [PubMed]

24. Lai, F.H.; Ng, D.S.; Bakthavatsalam, M.; Chan, V.C.; Young, A.L.; Luk, F.O.; Tsang, C.W.; Brelén, M.E. A multicenter study on the long-term outcomes of half-dose photodynamic therapy in chronic central serous chorioretinopathy. *Am. J. Ophthalmol.* **2016**, *170*, 91–99. [CrossRef] [PubMed]
25. Chhablani, J.; Cohen, F.B. Central Serous Chorioretinopathy International Group. Multimodal imaging-based central serous chorioretinopathy classification. *Ophthalmol. Retin.* **2020**, *4*, 1043–1046. [CrossRef]
26. Gawęcki, M.; Grzybowski, A.; Święch, A. Biometric risk factors for central serous chorioretinopathy. *Ophthalmol. Ther.* **2023**, *12*, 1327–1338. [CrossRef]
27. Gawęcki, M.; Grzybowski, A. Ganglion cell loss in the course of central serous chorioretinopathy. *Ophthalmol. Ther.* **2023**, *12*, 517–533. [CrossRef]
28. Yu, J.; Jiang, C.; Xu, G. Correlations between changes in photoreceptor layer and other clinical characteristics in central serous chorioretinopathy. *Retina* **2019**, *39*, 1110–1116. [CrossRef]
29. Torres-Costa, S.; Penas, S.; Cerqueira, A.R.; Brandão, E.; Carneiro, Â.; Rocha-Sousa, A.; Falcão-Reis, F. Long term outer retinal changes in central serous chorioretinopathy submitted to half-dose photodynamic therapy. *Photodiagn. Photodyn. Ther.* **2021**, *34*, 102235. [CrossRef]
30. Matsumoto, H.; Sato, T.; Kishi, S. Outer nuclear layer thickness at the fovea determines visual outcomes in resolved central serous chorioretinopathy. *Am. J. Ophthalmol.* **2009**, *148*, 105–110.e1. [CrossRef]
31. Kamimura, A.; Miki, A.; Kishi, M.; Okuda, M.; Hayashida-Hirano, M.; Sakamoto, M.; Matsumiya, W.; Imai, H.; Kusuhara, S.; Nakamura, M. Two-year outcome of half-time photodynamic therapy for chronic central serous chorioretinopathy with and without choroidal neovascularization. *PLoS ONE* **2023**, *18*, e0284979. [CrossRef] [PubMed]
32. Maruyama-Inoue, M.; Yanagi, Y.; Inoue, T.; Kitajima, Y.; Kadonosono, K. Importance of fluorescein angiography as a predictor of treatment response in patients with pachychoroid neovasculopathy. *Sci. Rep.* **2024**, *14*, 29157. [CrossRef] [PubMed]
33. Pauleikhoff, L.; Diederen, R.; Feenstra, H.; Schlingemann, R.; van Dijk, E.; Boon, C. Single-session bilateral reduced-settings photodynamic therapy for bilateral chronic central serous chorioretinopathy. *Retina* **2023**, *43*, 1356–1363. [CrossRef]
34. Khandhadia, S.; Thulasidharan, S.; Hoang, N.; Ibrahim, S.; Ouyang, Y.; Lotery, A. Real world outcomes of photodynamic therapy for chronic central serous chorioretinopathy. *Eye* **2023**, *37*, 2548–2553. [CrossRef] [PubMed]
35. Aisu, N.; Miyake, M.; Hosoda, Y.; Mori, Y.; Takahashi, A.; Muraoka, Y.; Ueda-Arakawa, N.; Miyata, M.; Oishi, A.; Tamura, H.; et al. Effectiveness of reduced-fluence photodynamic therapy for chronic central serous chorioretinopathy: A propensity score analysis. *Ophthalmol. Sci.* **2022**, *2*, 100152. [CrossRef]
36. Feenstra, H.; Diederen, R.; Lamme, M.; Tsonaka, R.; Fauser, S.; Yzer, S.; van Rijssen, T.; Gkika, T.; Downes, S.; Schlingemann, R.; et al. Increasing evidence for the safety of fovea-involving half-dose photodynamic therapy for chronic central serous chorioretinopathy. *Retina* **2023**, *43*, 379–388. [CrossRef]
37. Park, Y.J.; Kim, Y.K.; Park, K.H.; Woo, S.J. Long-term efficacy and safety of photodynamic therapy in patients with chronic central serous chorioretinopathy. *Ophthalmic Surg. Lasers Imaging Retin.* **2019**, *50*, 760–770. [CrossRef]
38. Wakatsuki, Y.; Tanaka, K.; Mori, R.; Furuya, K.; Kawamura, A.; Nakashizuka, H. Morphological changes and prognostic factors before and after photodynamic therapy for central serous chorioretinopathy. *Pharmaceuticals* **2021**, *14*, 53. [CrossRef]
39. Li, M.; Qu, J.; Liang, Z.; Tang, J.; Hu, J.; Yao, Y.; Jin, E.; Li, X.; Zhao, M. Risk factors of persistent subretinal fluid after half-dose photodynamic therapy for treatment-naive central serous chorioretinopathy. *Graefes Arch. Clin. Exp. Ophthalmol.* **2022**, *260*, 2175–2182. [CrossRef]
40. van Rijssen, T.J.; van Dijk, E.H.C.; Dijkman, G.; Boon, C.J.F. Clinical characteristics of chronic central serous chorioretinopathy patients with insufficient response to reduced-settings photodynamic therapy. *Graefes Arch. Clin. Exp. Ophthalmol.* **2018**, *256*, 1395–1402. [CrossRef]
41. Arrigo, A.; Calamuneri, A.; Aragona, E.; Bordato, A.; Grazioli Moretti, A.; Amato, A.; Bandello, F.; Battaglia Parodi, M. Structural OCT parameters associated with treatment response and macular neovascularization onset in central serous chorioretinopathy. *Ophthalmol. Ther.* **2021**, *10*, 289–298. [CrossRef] [PubMed]
42. Nakamura, K.; Takeuchi, J.; Kataoka, K.; Ota, H.; Asai, K.; Nakano, Y.; Horiguchi, E.; Taki, Y.; Ito, Y.; Terasaki, H.; et al. Effects of half-dose photodynamic therapy on chronic central serous chorioretinopathy with or without macular neovascularization assessed using optical coherence tomography angiography. *Retina* **2022**, *42*, 2346–2353. [CrossRef] [PubMed]
43. Mirshahi, R.; Naseripour, M.; Ghomashi, A.; Falavarjani, K.G. Clinical predictive factors and imaging biomarkers of treatment response to half dose PDT in patients with chronic central serous chorioretinopathy. *Photodiagn. Photodyn. Ther.* **2024**, *48*, 104224. [CrossRef]
44. Sousa, K.; Viana, A.R.; Pires, J.; Ferreira, C.; Queirós, L.; Falcão, M. Outer nuclear layer as the main predictor to anatomic response to half dose photodynamic therapy in chronic central serous retinopathy. *J. Ophthalmol.* **2019**, *2019*, 5859063. [CrossRef]
45. Son, K.Y.; Lim, S.G.; Hwang, S.; Choi, J.; Kim, S.J.; Kang, S.W. Foveal atrophy in patients with active central serous chorioretinopathy at first presentation: Characteristics and treatment outcomes. *Br. J. Ophthalmol.* **2024**, *109*, 89–97. [CrossRef]

46. Yu, J.; Ye, X.; Li, L.; Jiang, C.; Chang, Q.; Xu, G. Threshold thickness of foveal outer nuclear layer associated with outcomes of photodynamic therapy in central serous chorioretinopathy. *Eye* **2022**, *36*, 1884–1889. [CrossRef]
47. Jeong, S.; Kang, W.; Sagong, M. Influence of vortex vein engorgement for photodynamic therapy in central serous chorioretinopathy. *Sci. Rep.* **2024**, *14*, 24424. [CrossRef]
48. Ozkaya, A.; Garip, R.; Alkin, Z.; Taskapili, M. The comparison of multimodal imaging findings of central serous chorioretinopathy patients in regard to the early anatomically treatment response to half-fluence photodynamic therapy: A retrospective case-control study. *Int. J. Retin. Vitr.* **2017**, *3*, 20. [CrossRef]
49. Forte, P.; Cattaneo, J.; Cardillo Piccolino, F.; Arrigo, A.; Corazza, P.; Musetti, D.; Rosa, R.; Traverso, C.E.; Fontana, V.; Lupidi, M.; et al. Influence of scleral thickness on photodynamic therapy outcomes in central serous chorioretinopathy. *Acta Ophthalmol.* **2024**, *103*, e165–e175. [CrossRef]
50. Lim, J.I.; Glassman, A.R.; Aiello, L.P.; Chakravarthy, U.; Flaxel, C.J.; Spaide, R.F. Macula Society CSC Collaborative Study Group, Research and Education Committee and Website Committee. Collaborative retrospective macula society study of photodynamic therapy for chronic central serous chorioretinopathy. *Ophthalmology* **2014**, *121*, 1073–1078. [CrossRef]

MDPI AG
Grosspeteranlage 5
4052 Basel
Switzerland
Tel.: +41 61 683 77 34

Biomedicines Editorial Office
E-mail: biomedicines@mdpi.com
www.mdpi.com/journal/biomedicines

www.ingramcontent.com/pod-product-compliance
Lightning Source LLC
Chambersburg PA
CBHW062059220926
51850CB00010B/2821